AF577343

5 063054 01

THE

OXFORD ENGINEERING SCIENCE SERIES

General Editors

L. C. WOODS, W. H. WITTRICK, A. L. CULLEN

SYNTHESIS OF PLANAR ANTENNA SOURCES

BY

DONALD R. RHODES

UNIVERSITY PROFESSOR
NORTH CAROLINA STATE UNIVERSITY

CLARENDON PRESS · OXFORD
1974

Oxford University Press, Ely House, London W. 1

GLASGOW NEW YORK TORONTO MELBOURNE WELLINGTON
CAPE TOWN IBADAN NAIROBI DAR ES SALAAM LUSAKA ADDIS ABABA
DELHI BOMBAY CALCUTTA MADRAS KARACHI LAHORE DACCA
KUALA LUMPUR SINGAPORE HONG KONG TOKYO

ISBN 0 19 856123 7

PRINTED IN NORTHERN IRELAND AT
THE UNIVERSITIES PRESS BELFAST

TO

ROGER C. RHODES

whose emerging career
I eagerly await

The bitter and the sweet come from the outside,
the hard from within, from one's own efforts.

ALBERT EINSTEIN
Self-portrait

PREFACE

Occasionally one experiences an event that proves to be a turning point in the course of one's life. It was just such an event that led ultimately to the development of this treatise, so perhaps the reader will forgive a few personal remarks pertaining to it and to the circumstances under which it occurred. It happened a little over a decade ago on the 16th of August 1962 when I first read Slepian and Pollak's recently published paper on the double orthogonality properties of the prolate spheroidal wave functions of order zero. A decade before that, in 1952, I had spent a full year on an unsuccessful attempt to formulate and solve a problem of antenna pattern shaping by synthesizing an optimum line source. Although the attempt itself had met with total failure it left a deep and lasting conviction that no truly scientific basis for antenna design would ever be developed without first developing an understanding of the physical and mathematical problems that are peculiar to synthesis. With no prospects in sight for acquiring such an understanding I left the field of antenna theory completely until the day that Slepian and Pollak's paper came to my attention. Here for the first time was the kind of functions needed to solve the synthesis problem on which I had laboured so unsuccessfully ten years earlier. Once the double orthogonality properties were recognized the solution was almost obvious, and by the end of the day it was finished. In July 1963 it was published in what was to become the first of fourteen papers on various investigations into the synthesis of planar antenna sources. The present treatise is an attempt to bring together the principal results of those investigations, including some that have not been published previously.

Although that first solution contained the most essential elements of optimum source synthesis it soon became evident that its physical meaning was not entirely clear. This led to the decision to embark upon a systematic programme of investigation into the theoretical foundations of antenna synthesis. As nothing of this nature had been attempted before it meant starting out on a search for the conditions that must be satisfied by physically realizable synthesis solutions. Within a year the first result was obtained that set the stage for much of the later work. This was the formulation of the fundamental principle of planar sources developed in Chapter 3. An immediate consequence of the new principle was an analytic extension of the classical Poynting vector method commonly used for calculating the radiation resistance of an antenna from its radiation pattern. By extending the Poynting integration of the pattern space factor from its visible region into the invisible region obtained by analytic continuation it was found that the input reactance of the antenna could be calculated by exactly the same process as that used to obtain radiation resistance. A second consequence was obtained two years later when the concept of observable stored energies was formulated together with explicit expressions for the observable stored magnetic and electric energies of planar sources. This concept of observable stored energies, although not yet fully developed even today, appears to be essential for any Q-constrained synthesis solution because of the fact that it provides the only formulation for the quality factor Q that is internally consistent with its presumed relationship to relative input bandwidth.

The first six chapters are concerned largely with a development of the mathematical conditions that must be satisfied for physical realizability and of mathematical methods for satisfying those conditions, while the last three chapters describe actual synthesis methods for the three different types of pattern goodness criteria introduced in Chapter 1. A detailed treatment of the properties of rectangular and circular sources is developed in Chapter 4. Such two-dimensional sources are shown to be separable physically into the product of an edge factor and an aperture space factor, somewhat analogous to the way in which their radiation pattern always separates into an element factor and a pattern space factor. The ensuing relationship between the aperture space factor and pattern space factor then leads to the existence of certain functions that are characteristic of all rectangular sources and to certain other functions that are characteristic of all circularly symmetric sources. In the case of rectangular sources these characteristic functions consist of products of spheroidal functions, while for circularly symmetric sources they come from a generalization of the spheroidal functions. But the only planar sources for which a fully workable mathematical formulation has yet been found for solving synthesis problems of practical interest are strip and line sources. The relevant properties of strip and line sources are developed in Chapter 5. Two aspects of these essentially one-dimensional planar sources have proved to be particularly important. One is the fact that their characteristic functions, for any given value of the edge exponent α that determines the edge factor of the source, are simply the spheroidal functions of order α. This prompted a number of purely mathematical investigations into the theory of spheroidal functions, leading to the discovery of some new properties of the functions themselves. The most important of those properties are summarized in Appendix 3. It may be of interest to note here that one of them, the existence of a third kind of characteristic numbers, appears to be unique in the annals of special function theory. The second aspect of importance was the formulation of a constraining parameter γ_α that, for certain values of α, was found to be related either exactly or approximately to Q and at the same time to take full advantage of the double orthogonality properties of the charerteristic spheroidal functions of order α.

A theory of pattern sampling, as an alternative to the theory of orthogonal functions for constructing the source required to produce an arbitrary aperture-limited pattern space factor, is presented in Chapter 6. It begins with a description of the conventional Woodward sampling method and concludes with the development of a completely general theory of sampling synthesis.

Explicit synthesis solutions for maximum resolution, for maximum directivity, and for minimum pattern-shaping error are developed in the last three chapters. Taylor's near-optimum solution to the maximum resolution problem is generalized in Chapter 7 to admit an arbitrary choice of the edge exponent α. This is followed in Chapter 8 by the development of a class of optimum solutions to the maximum directivity problem, and in Chapter 9 by the development of a class of optimum solutions to the pattern-shaping problem. The latter is a generalization of the solution obtained ten years ago to the pattern-shaping problem that originally motivated this series of investigations. These two optimum solutions possess two distinguishing properties that are of fundamental importance to the theory of source synthesis. One is the fact that they have a clear physical meaning that the earlier pattern-shaping solution did not. The other is the fact that they give the best possible value for their pattern goodness measures under the constraining conditions specified. It is impossible to do better without relaxing the constraints.

No attempt has been made to include in this treatise all of the many results that have been published by others on the problem of source synthesis. The few that have been included are limited largely to those that have had a direct bearing upon my own researches, either as a stimulus to them or as an outgrowth of them. In that sense, then, the body of theory as presented herein can by no means be considered complete.

A long-range programme such as this would have been extremely difficult to carry through without considerable support. Dr. Alton W. Sissom saw to it that Radiation Incorporated, of which I was a member of the technical staff during the early years, provided the support needed to get the programme started. Dr. Allan C. Schell became interested in the programme in the summer of 1963, and shortly thereafter he and Mr. Carlyle J. Sletten arranged to have the Air Force Cambridge Research Laboratories assume a large part of the financial burden. Their support, and their sympathetic understanding at times when the going was difficult, continued almost without interruption for the next eight years. In 1966 I joined the faculty of the North Carolina State University where Dr. George B. Hoadley, Head of the Electrical Engineering Department, provided an academic environment that proved to be conducive to an intensive research activity of this kind. To all four men and to the organizations they represented I extend my sincerest thanks and warm appreciation. I would like also to acknowledge my appreciation to a very capable typist, Miss Sandra M. Maxim, for her patience in keeping up with the constantly changing manuscripts that have come out of the programme.

DONALD R. RHODES

Raleigh, North Carolina
May 1973

CONTENTS

CHAPTER 1
PHYSICAL NATURE OF ANTENNA SYNTHESIS

THE engineer who designs antennas for a living must surely consider antenna theorists to be a peculiar lot. He is a practical man who is concerned with practical things like system specifications, and he knows from experience that some specifications may be incompatible with others; e.g. he knows that he can never build an antenna whose beamwidth is too small, or whose sidelobes are too low, for the size of aperture available. But when he turns to an antenna theorist to find out the extent to which Nature has conspired against him and what design compromises might be available to him he finds that they don't even speak the same language. Instead of system specifications the theorist wants to talk in terms of particular configurations of dipoles, loops, slots, horns, and so on because these are the kinds of things that he knows how to handle. Once the antenna configuration has been specified fully he feels confident that he can determine all of its performance characteristics, including things like the radiation pattern, the input impedance, and the input bandwidth. But things like system specifications make him uncomfortable because he's not quite sure what to do with them.

Therein lies the difference between the practicing engineer and the theorists to whom he turns for help. And it's not just a difference of language. The engineer has come face to face with the subtle and difficult problems of synthesis, whereas antenna theorists have elected, by and large, to remain in the more comfortable and friendly world of analysis. As a consequence the design methods available to the engineer are limited almost exclusively to those of analysis rather than synthesis. First he must hypothesize a particular antenna structure, based largely upon past experience. Then he must analyse it to determine its performance. If his first guess proves to be inadequate he can change the design, again on the basis of past experience, and then redetermine its performance. The closeness with which his final design meets the specifications given will thus be heavily dependent upon experience and judgement. And when he is done he has no way of knowing how close he has come to achieving the best design possible.

The ultimate objective of antenna synthesis is to replace this empirical approach by a systematic procedure that leads directly to the best design permitted by the specifications given. But synthesis problems are of an essentially different character than problems of analysis. Antenna analysis is straightforward, in that the performance of any antenna whose structure is specified in advance can always be determined exactly and completely, in principle if not in practice, by solving the appropriate boundary-value problem. Antenna synthesis, on the other hand, is not at all straightforward even in principle. In fact, there is no *a priori* assurance that the system specifications given can be satisfied at all by a real antenna. Worse yet, the specifications themselves may even be mutually incompatible. Thus, the familiar problems of analysis, for which exact physical solutions always exist, may bear little resemblance to the relatively unfamiliar problems of synthesis, for which there may be no physical solution of any kind.

The most essential difference between analysis and synthesis lies in the question of physical realizability. Solutions to problems of analysis are inherently realizable if no approximations are made in their formulation. But realizable solutions to problems of synthesis, if they exist at all, can be guaranteed only by building the conditions for physical realizability directly into the mathematical description of the problem itself. In this respect antenna synthesis is somewhat similar to network synthesis; the system spectral (frequency) functions representing voltages, currents, or fields of an antenna must be constructed in such a way that they satisfy precisely the same realizability condition of analyticity in the lower half of the complex frequency plane as must the system spectral functions of linear networks. But here the similarity stops. Whereas network synthesis involves only the frequency variable, antenna synthesis involves the spatial variables as well. The spatial variables complicate the problem of physical realizability considerably beyond that of network synthesis because they introduce additional restrictions that appear to have no counterpart in network synthesis. Three additional realizability conditions are now known in the particular case of planar antennas: aperture-limitation, which results in analyticity of the pattern space factor; aperture edge behavior, which determines the asymptotic behaviour of the pattern space factor; and a constraint on the proportion of reactive to radiative power at the input port of the antenna, which limits the size of the invisible lobes of the pattern space factor. These three realizability conditions appear to be at the very heart of the antenna synthesis problem. The way in which they arise, and some of their consequences in antenna design, are the principal subjects of this treatise.

1.1. System specifications

When an antenna is to be designed as one part of a complete electronic system the specifications given to the designer are usually in the form of certain operational requirements on the system as a whole: size, weight, performance characteristics, and so on. Consider, for example, the case of an antenna that is to be designed as part of an air-to-ground search radar in a supersonic aircraft. There will be certain spatial limitations on the size and shape of the surface area available for the radiating aperture, also on the size and shape of the volume available for the antenna as a whole, that will be imposed by the design of the aircraft structure. When these spatial limitations are expressed in terms of electrical dimensions at the operating frequency specified they, in turn, will lead to certain fundamental design limitations upon the performance available. Thus, some of the performance specifications given will almost surely have to be compromised.

One performance specification that almost surely must be compromised is the radiation pattern. Actual radiation patterns of physical antennas will be shown in section 3.1 to constitute a very special class of analytic functions that excludes virtually all of the idealized pattern functions usually specified. In the case of air-to-ground search radars the radiation pattern ideally should have a narrow beam in the horizontal plane and a cosecant-shaped beam in the vertical plane, with no sidelobes in either plane. But such functions cannot be realized exactly, not even in principle. They can only be approximated. Thus, one or more pattern goodness criteria by which quantitative measures of the accuracy of the pattern approximation can be obtained must also be specified. Two different criteria are usually used simultaneously in the case of search radar antennas, namely, maximum resolution in the horizontal plane and minimum pattern-shaping error, in some specified sense, from the cosecant-shaped beam that is desired in the vertical plane.

There would be no limit to the pattern accuracy available if no constraints were imposed upon the synthesis solution other than the limitation on electrical dimensions of the aperture area and antenna volume. For an aperture of any given size one can always calculate an aperture distribution whose radiation pattern is arbitrarily close to any function desired. This is the famous result of Bouwkamp and de Bruijn† that suggested the theoretical possibility of achieving superdirective aperture distributions. But if the actual pattern were forced to approach the unrealizable pattern specified over the visible region of radiative power then its analytic continuation into the invisible region of reactive power would have to grow without bound. Hence the observable stored magnetic and electric energies, defined later in section 3.4 and shown in section 3.5 to increase as the square of the magnitude of the pattern space factor in its invisible region, would also have to grow without bound. And so would the quality factor Q, which will be seen to be proportional to the total observable stored energy. If one makes the usual assumption that the half-power bandwidth seen at the input port of the antenna approaches the reciprocal of Q asymptotically with increasing values of Q, then an unbounded value of Q would correspond to a vanishingly small value of the input bandwidth. This, in turn, would

† C. J. Bouwkamp and N. G. Bruijn. The problem of optimum antenna current distribution. *Philips Research Reports*, **1**, pp. 135–58, January 1946.

limit the use of such antennas to signals whose bandwidth is vanishingly small. As a consequence the capacity of the channel to communicate information, shown by Shannon† to be proportional to signal bandwidth for signals of given average power when perturbed by white thermal noise of given average power, would also have to vanish. Thus, even if antennas with arbitrarily high directivity could be constructed physically they would be of no practical use for communicating information.

It is evident, then, that some additional constraints must be imposed upon the synthesis solution in order to ensure physical realizability. The most important of these is an upper bound on Q, in order to ensure at least a rough lower bound on input bandwidth.

It would be quite possible, of course, to maintain the input bandwidth at any specified value simply by building enough heat dissipation into the antenna itself to compensate for the tremendous growth in reactive power. But the power dissipated in heat would then have to approach the same enormous level as the reactive power, which would cause the radiation efficiency to fall so low that the antenna would again become useless. Hence, a second performance parameter on which some form of realizability constraint must be imposed, either explicitly or implicitly, is the radiation efficiency of the antenna.

Still a third performance parameter on which a constraint must be imposed is the input impedance. This is a consequence of the fact that any significant deviation of the input impedance from the value required to match a transmission line would require that some matching network be included. The matching network would then become, in effect, a part of the antenna system. It would have the deleterious effect of introducing additional reactances whose observable stored energies would act to increase the Q of the system from its specified value.

In summary, then, the system specifications that determine the electrical design of the antenna appear to fall into four different categories:

(a) electrical dimensions of the aperture area and of the antenna volume available;
(b) radiation pattern desired;
(c) pattern goodness criteria;
(d) performance parameters, including Q or input bandwidth, radiation efficiency, and input impedance.

The object of the antenna synthesis problem is to find the internal electrical structure of the antenna that produces the closest approximation to (*b*) that is permitted by (*a*), (*c*), and (*d*).

† C. E. Shannon. The mathematical theory of communication. *Bell System Tech. J.*, **27**, pp. 379–423 and 623–656, July and October 1948; Theorem 17. Reprinted by University of Illinois Press, Urbana, Illinois, 1959.

1.2. Pattern-goodness criteria

There are three different types of pattern goodness criteria in common use. Two of them—maximum resolution and maximum directivity—apply only to narrow beam patterns. The third—minimum pattern-shaping error, in some specified sense, over the visible region—applies to any type of shaped-beam pattern. Minimum mean-squared error is a common example of one sense in which a pattern-shaping error criterion might be specified. The particular criterion, or combination of criteria, specified for any particular antenna design would be determined ideally by the operational requirements on the system. Each of the three different types of criteria will be described briefly here. Syntheses obtainable for specific examples of each of the three will be treated in detail in Chapters 7, 8, and 9.

(i) *Maximum resolution*

For receiving antennas in which the design objective is to discriminate against all signals arriving from directions other than some specified direction the radiation pattern should be in the form of a single narrow beam that is shaped to give maximum resolution. A common measure of resolution is the half-power beamwidth; i.e. the angular width between the directions in which the radiated power density drops to half of its maximum value. This is very nearly the classical Rayleigh criterion for resolution that is commonly used in optics, where little control over pattern shape is available. But it is not as sharp a criterion as required for antenna design because it imposes no restrictions on the relative signal strength permitted outside the main beam. It can be made sharper by including a constraint on the permissible sidelobe ratio. Thus, the pattern-goodness criterion for maximum resolution will be taken here to be minimum beamwidth for a prescribed sidelobe ratio.

(ii) *Maximum directivity*

When the design objective is to obtain the greatest concentration of power in a single direction the radiation pattern should again be made to come as close as possible to having a single narrow beam. But the shape of the narrow beam pattern need not, and usually will not, be the same as for the maximum resolution criterion. Here the appropriate criterion is maximum directivity. No constraining condition on sidelobe ratio is necessary because the presence of sidelobes is no longer a matter of concern. The sole matter of concern is to maximize the power density radiated in a particular direction relative to the average power radiated per unit angle.

(iii) *Minimum pattern-shaping error*

A quite different kind of criterion is required when the objective is to obtain a best approximation to the shape of some desired radiation pattern.

The desired, but unrealizable, radiation pattern may be of any shape whatever. Hence the pattern goodness criterion must be minimum pattern-shaping error over some or all of the visible region. One must then specify, in addition, the *sense* in which the pattern-shaping error is to be determined. There is an unlimited number of possible senses that one might choose, each leading to a different error measure. The simplest is to force the pattern approximation to fit the ideal pattern exactly at some discrete set of points within the visible region, as described in Chapter 6. This has the disadvantage, however, that it provides no control over the accuracy of approximation within the intervals between those points. A quite different one is to form some kind of weighted mean-square error over the entire visible region, an example of which will be treated in detail in Chapter 9. By choosing the weight factor properly this example will be seen to result in minimum radiated power in the error pattern.

1.3. Antenna synthesis

Once the system specifications have been given in a mutually compatible form it should be possible, in principle, to deduce the properties of the electrical medium that must be placed within the volume available in order to obtain the best possible values of the pattern goodness measures chosen. The resulting description of the electrical medium would then be the optimum physically realizable solution sought.

The electrical properties of all linear material media are known to be describable completely in terms of their complex electric permittivity, frequently expressed in terms of the dielectric constant ϵ and conductivity σ, and their complex magnetic permeability μ. All three of these parameters will be functions of the three position coordinates x, y, z and of frequency ω, in principle, the latter from the fact that all physical media must be dispersive. Hence the ultimate objective of antenna synthesis, in its most fundamental form, might be formulated somewhat as follows. It is to determine the constitutive functions $\epsilon(x, y, z, \omega)$, $\sigma(x, y, z, \omega)$, and $\mu(x, y, z, \omega)$ that would be required within a closed surface of specified dimensions in order to obtain the best possible values of the pattern goodness measures for arbitrarily specified values of each of the constraining performance parameters.

Although this formulation of the antenna synthesis problem is totally different from the empirical formulation usually used in antenna design it is conceivable, of course, that the two could lead to similar physical configurations. But it is highly unlikely that the optimum solution obtained in this way for ϵ, σ, and μ in the spatial and frequency domains would just happen to coincide with an empirically selected collection of dipoles, loops, and so on. On the other hand, the differences between the performance of the optimum solution and some judiciously chosen empirical configuration could turn out to be relatively unimportant. In such a case one might conclude that the effort expended in finding the optimum solution was wasted. But to do so would be to miss the point. The point is that the optimum solution establishes the absolute upper bound on the performance available, beyond which no physical solution of any kind is possible. Thus, even if the optimum solution itself were never implemented it still would have served the useful function of establishing the closeness with which the actual implementation, regardless of how it may have been obtained, comes to giving the best performance possible.

Actual implementation of the optimum solution becomes a materials problem that is not properly within the domain of antenna synthesis itself. It is an independent problem of constructing a material medium whose complex electric permittivity and magnetic permeability varies with position and with frequency in the manner required. A wide variety of methods are available, including the use of ordinary dielectrics whose electrical properties are controlled in the normal process of manufacture and artificial dielectrics whose electrical properties are controlled by loading ordinary dielectrics with small conducting objects. Both methods have been used extensively in microwave lens design.†

1.4. Source synthesis

The antenna synthesis problem as outlined above is presented here only as a goal for future research. It cannot be solved as yet, not only because of computational difficulties but because it involves physical concepts that are still not completely understood. In the particular case of antennas with planar apertures, however, some of the most essential concepts have been clarified, at least in part. It is known, for example, that the pattern space factor is an analytic function of its space and frequency variables and that it bears a unique reversible relationship with the source (the aperture distribution) through a Fourier integral transformation. From this it has been found that the concept of energy storage in the fields of planar sources can be made logically consistent with its role in determining input bandwidth by introducing the new concept of observable stored energies. But these concepts pertain only to the volume outside the antenna. Nothing at all is known as yet about the volume inside. For these reasons the body of theory developed in this treatise has been limited exclusively to the synthesis of planar sources. By assuming that the material medium inside the antenna is lossless, and that the reactive effects of energy stored within it can be approximated by just a single lumped reactance, the synthesis problem for planar antennas can be reduced to a determination of just the source alone.

† S. A. Schelkunoff and H. T. Friis. *Antennas—theory and practice*, John Wiley, New York, 1952; pp. 573–591.

It is important to recognize, however, that source synthesis is only a substitute for, and an approximation to, the actual problem of antenna synthesis. The principal justification for it is that it provides explicit examples of some of the properties and limitations to be expected of antenna synthesis itself. But as an approximation to the actual problem of antenna synthesis its accuracy is still unknown. And it provides no knowledge at all as to how the required antenna configuration should be constructed. All that it provides is an aperture distribution, not a description for the antenna as a whole.

The subject of source synthesis is frequently, but erroneously in the opinion of the author, referred to as pattern synthesis. The reason why I believe it to be erroneous is that it places the emphasis upon the wrong part of the problem. The objective is not to synthesize the radiation pattern specified, which isn't even realizable in most practical cases of interest, but rather to synthesize the source that produces some best approximation to it. It is the source, not the radiation pattern, that is under the most direct control of the antenna design engineer. Hence it is the source, not the radiation pattern, that the engineer seeks to synthesize by means of actual hardware.

CHAPTER 2
FIELDS OF PLANAR SOURCES

ANY mathematical formulation for a physically realizable theory of antenna synthesis must be expressed ultimately in terms of fields that satisfy Maxwell's equations and the boundary conditions. By assuming that the antenna has a perfectly conducting outer surface containing a radiating aperture, the fields everywhere outside of the antenna will be determined completely and uniquely by just the tangential electric field in the aperture. For time-harmonic fields the tangential electric field in the aperture is usually referred to as the *aperture distribution* or, more simply, as the *source*. In the case of planar antennas the source has the immensely important property of being determined exactly and uniquely by its radiation pattern. This is a consequence of the fundamental principle of planar sources that will be established in Chapter 3 and treated there in some detail. In succeeding chapters the principle will be seen to play a key role in the synthesis of planar sources because of the fact that it places the burden of the pattern approximation directly upon the radiation pattern itself rather than indirectly through the source. Nothing like it has yet been established for arbitrary nonplanar sources. This is the principal reason why the researches on which this treatise is based were limited solely to sources that are planar.

By a *planar antenna* is meant specifically any antenna whose radiating aperture consists of one or more holes in a perfectly conducting planar sheet of infinite extent. In its most elementary form it need be no more than a simple slot of arbitrary size and shape cut in a thin conducting planar sheet (a slot antenna) that is excited by means of a probe inserted into the slot and that is free to radiate equally on both sides of the sheet. An equally elementary form is that of the complementary antenna (a dipole antenna) obtained by interchanging the conducting and nonconducting parts of the sheet. A far more versatile form, however, is one in which the radiating aperture is excited by means of a conducting cavity that completely surrounds the aperture on one side of the conducting sheet. Radiation from such an aperture is then limited to one side of the sheet, which means that no complementary form is possible. These three are the only distinctly different forms of planar antenna possible. Planar arrays can be constructed in any one of the three forms.

This definition for what is meant by a planar antenna constitutes a model involving two idealizations that can never be realized exactly in practice, namely, that the planar sheet is (*i*) perfectly conducting and (*ii*) infinite in extent. But the consequences of both idealizations can usually be approximated rather closely in practice by means of a copper sheet that extends a few wavelengths beyond the aperture. Thus, both idealizations appear to provide a realistic model for many of the antennas used in actual practice.

An exact formulation for the fields produced by planar sources of arbitrary size and shape will be developed in this chapter. It is a simple extension of the classic analysis of one-dimensional transverse-magnetic sources by Woodward and Lawson† to the case of completely general two-dimensional planar sources. The fields will be expressed in terms of two scalar functions that are interpretable as rectangular components of the vector pattern space factor. At any given frequency these two components are functions of the direction cosines of any point in the far field. For all real angles the pattern space factor will be shown to determine the radiation pattern; all the rest will be seen in the next chapter to determine the reactive properties of the source. Next, the connection between the properties of complementary planar antennas (slots and dipoles) will be established, including Booker's relationship between their input impedances. Finally, the behaviour that all physically realizable planar sources must have at their edges will be derived. This leads to a more realistic model than that which is usually assumed for planar slot and dipole antennas.

2.1. Construction of the field functions

For any linear, time-invariant, passive electromagnetic system, whether radiating or not, the fields $\mathscr{E}(x,y,z,t)$ and $\mathscr{H}(x,y,z,t)$ produced by an arbitrary driving voltage $v(t')$ at the input port can be represented exactly in terms of impulse responses of the system by

$$\mathscr{E}(x,y,z,t)=\int\limits_0^{t-\frac{r}{c}} v(t')\hat{\mathscr{E}}(x,y,z,t-t')\,dt', \tag{2.1a}$$

$$\mathscr{H}(x,y,z,t)=\int\limits_0^{t-\frac{r}{c}} v(t')\hat{\mathscr{H}}(x,y,z,t-t')\,dt', \tag{2.1b}$$

where r represents the distance $(x^2+y^2+z^2)^{\frac{1}{2}}$ from the input port and c denotes the velocity of light in free space. The functions $\hat{\mathscr{E}}(x,y,z,t-t')$ and $\hat{\mathscr{H}}(x,y,z,t-t')$ represent, respectively, the electric and magnetic field responses that would be produced at the spatial point (x,y,z) at time t by a unit impulse of voltage at the input port $(0,0,0)$ of the system at time t'. Their purpose is to provide a mathematical description of the system

† P. M. Woodward and J. D. Lawson. The theoretical precision with which an arbitrary radiation pattern may be obtained from a source of finite size. *J. Instn elect. Engrs*, Part III, **95**, pp. 363–70, September 1948.

alone, independently of the particular input signal $v(t')$ that might be used to drive it.

An equivalent mathematical description of the system alone, and the one that will be used throughout this treatise because of certain essential simplifications that will occur in the ensuing development of the theory, is the Fourier transform of the impulse–response field functions,†

$$\mathbf{E}(x, y, z, \omega) = \int_{-\infty}^{\infty} \hat{\mathscr{E}}(x, y, z, \tau)e^{-i\omega\tau}\, d\tau, \tag{2.2a}$$

$$\mathbf{H}(x, y, z, \omega) = \int_{-\infty}^{\infty} \hat{\mathscr{H}}(x, y, z, \tau)e^{-i\omega\tau}\, d\tau, \tag{2.2b}$$

where τ denotes $t-t'$ and i denotes $\sqrt{(-1)}$. The time-domain functions $\hat{\mathscr{E}}(x, y, z, \tau)$ and $\hat{\mathscr{H}}(x, y, z, \tau)$ represent actual fields in the system at any point (x, y, z) in space, hence they are always real for real values of the time variable τ. Their mathematically equivalent frequency-domain functions $\mathbf{E}(x, y, z, \omega)$ and $\mathbf{H}(x, y, z, \omega)$, on the other hand, do not of themselves represent actual fields. They are complex, in general, for both real and complex values of the frequency‡ variable ω. For real values of ω, and only for real values, they are the familiar phasor functions on which all of time-harmonic field theory is usually based.§ But it is important to recognize that it is this definition (2.2) in terms of impulse responses of the system, not their usual definition as phasors in time-harmonic field theory, that is the origin of the essential property of analyticity possessed by these frequency-domain functions in the lower half of their complex ω-plane. Furthermore their physical significance lies not in the fact that they can be used to describe time-harmonic fields, since any signals associated with time-harmonic fields would have zero bandwidth and hence would be totally incapable of conveying information, but rather in the fact that they provide a complete mathematical description of the particular system in which those fields must exist. Hence, instead of the name complex electric and magnetic fields usually given to them it would be more appropriate to call them the system

† The negative sign in the exponent of the complex exponential $e^{-i\omega\tau}$ is purely a matter of choice. This particular choice has the advantage that it leads to the conventional engineering expression $V_0 \operatorname{Re}\{\mathbf{E}(x, y, z, \omega_0)e^{i\omega_0 t}\}$ for the time-harmonic electric field produced at (x, y, z, t) by a sinusoidally varying input voltage of real angular frequency ω_0 and amplitude V_0. Although some authors (principally physicists) choose the opposite sign, such a choice would have the distinct disadvantage that positive reactances would have to be interpreted as capacitive and negative reactances as inductive, in opposition to the convention universally adopted throughout electrical engineering.

‡ Strictly speaking, the usual interpretation of ω as an angular frequency is valid physically for positive real values only.

§ For an example of the way in which phasors are usually introduced into time-harmonic field theory see R. F. Harrington, *Time-harmonic field theory*, McGraw-Hill, New York, 1961; p. 15.

functions for electric and magnetic field. For the sake of brevity, however, as well as to avoid too sharp a break with tradition, they will be referred to henceforth as simply the *field functions*.

The objective here is to construct the field functions that describe planar antenna systems in their exterior region only. It will be assumed that the antenna radiates into free space through an aperture A cut in a perfectly conducting planar sheet in the plane $z = 0$, as in Fig. 2.1. The aperture may consist of a single hole (a slot, for example) or any number of holes (an array of slots) of arbitrary size and shape. The field functions in the exterior

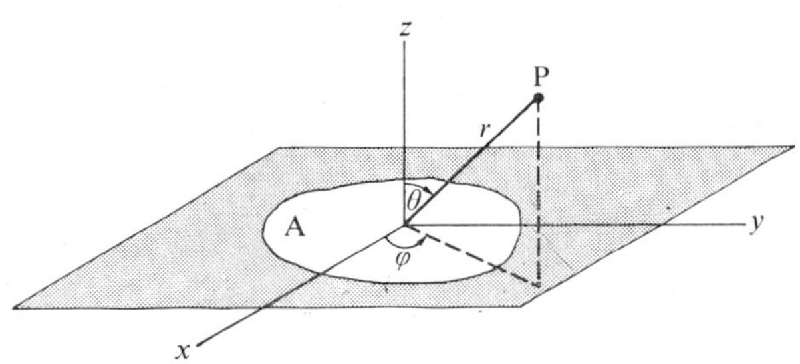

FIG. 2.1. Coordinate system for an aperture A in a conducting planar sheet.

half-space $z > 0$ must satisfy the *system equations*,

$$\nabla \times \mathbf{E}(x, y, z, \omega) = -i\omega\mu_0 \mathbf{H}(x, y, z, \omega), \tag{2.3a}$$

$$\nabla \times \mathbf{H}(x, y, z, \omega) = i\omega\epsilon_0 \mathbf{E}(x, y, z, \omega), \tag{2.3b}$$

obtained by taking the Fourier transform of Maxwell's equations for the impulse-response fields $\hat{\mathscr{E}}(x, y, z, \tau)$ and $\hat{\mathscr{H}}(x, y, z, \tau)$ that describe the system in the time domain. And they must also satisfy the boundary conditions on tangential E at the aperture plane $z = 0$ and the radiation condition at infinity. On the conducting part of the aperture plane the tangential components of $\mathbf{E}(x, y, z, \omega)$ must vanish, since the tangential components of the impulse-response field $\hat{\mathscr{E}}(x, y, z, \tau)$ must vanish there, from which it follows that the field functions everywhere in the half-space $z > 0$ must be determined uniquely by tangential E over just the aperture alone. The distribution of $E_{\tan}(x, y, z, \omega)$ over the aperture is what is meant by the source.

From this pair of partial differential equations one can obtain a wave equation for each of the rectangular components of $\mathbf{E}(x, y, z, \omega)$ by taking the curl of (2.3*a*), and for each of the rectangular components of $\mathbf{H}(x, y, z, \omega)$ by taking the curl of (2.3*b*). Thus, all six of the electric and magnetic field components must satisfy the scalar wave equation,

$$\nabla^2\Phi + k^2\Phi = 0, \tag{2.4}$$

where Φ denotes any one of the rectangular components of $\mathbf{E}(x, y, z, \omega)$ or $\mathbf{H}(x, y, z, \omega)$ and where k denotes the wavenumber, or scalar propagation constant, $\omega(\mu_0\epsilon_0)^{\frac{1}{2}}$ of free space.

Solving (2.4) by separation of variables one obtains elementary solutions of the form

$$\Phi = Fe^{i(\pm k_x x \pm k_y y \pm k_z z)}, \tag{2.5}$$

with arbitrary complex amplitudes F. The three separation parameters are complex numbers $-k_x^2$, $-k_y^2$, and $-k_z^2$ that are related to k^2 by

$$k_x^2+k_y^2+k_z^2 = k^2 \tag{2.6}$$

but that are arbitrary otherwise. Any two of the three separation parameters can be considered to be independent variables ($-k_y^2$ and $-k_z^2$ were chosen in [2]†), but the most convenient choice is $-k_x^2$ and $-k_y^2$. Hence the third, k_z, is determined by those two and by k,

$$k_z = (k^2-k_x^2-k_y^2)^{\frac{1}{2}}, \tag{2.7}$$

except for the question of sign.

The most general solution of (2.4) will be a linear combination of all physically admissible elementary solutions,‡

$$\Phi(x, y, z, k) = \frac{1}{(2\pi)^2} \iint_{C_{k_x} C_{k_y}} F(k_x, k_y, k) \times$$

$$\times \exp[i\{\pm k_x x \pm k_y y \pm (k^2-k_x^2-k_y^2)^{\frac{1}{2}} z\}]\, dk_x\, dk_y, \tag{2.8}$$

in which the unknown complex amplitude function $F(k_x, k_y, k)$ and the integration contours C_{k_x} and C_{k_y} in the complex k_x- and k_y-planes, respectively, are to be determined from the boundary conditions and the radiation condition. The radiation condition requires that the field functions for real values of k represent outgoing waves of decreasing magnitude at increasing distance from the aperture. If either k_x or k_y were permitted to have an imaginary part the integral (2.8) would grow without bound as x or y, respectively, approaches $\pm\infty$. Hence it is concluded that both k_x and k_y must have no imaginary part. This means that the contours C_{k_x} and C_{k_y} must each be restricted to the real axis; i.e. k_x and k_y must both be purely real. Furthermore, for the field functions to represent outgoing rather than incoming waves the signs in the exponentials of (2.8) must all be negative, as will be seen later in section 2.4. Thus, for the radiation condition to be

† All numbered references are to papers by the author, listed on p. 201.

‡ Henceforth the complex frequency variable will be taken to be the wavenumber k instead of the angular frequency ω, as a matter of convenience.

satisfied all six of the field components must be of the form

$$\Phi(x, y, z, k) = \frac{1}{(2\pi)^2}\int_{-\infty}^{\infty}\int_{-\infty}^{\infty} F(k_x, k_y, k)\times$$
$$\times \exp[-i\{k_x x+k_y y+(k^2-k_x^{\,2}-k_y^{\,2})^{\frac{1}{2}}z\}]\, dk_x\, dk_y, \quad (2.9)$$

in which the imaginary part of $(k^2-k_x^{\,2}-k_y^{\,2})^{\frac{1}{2}}$ must always be taken to be negative,

$$(k^2-k_x^{\,2}-k_y^{\,2})^{\frac{1}{2}} = -i(k_x^{\,2}+k_y^{\,2}-k^2)^{\frac{1}{2}}, \quad (2.10)$$

in order to prevent the integral from becoming unbounded as z approaches $+\infty$. This leaves only the six complex amplitude functions still undetermined. It will now be shown that each of these six functions is determined exactly and uniquely by the source alone.

For the three rectangular components of $\mathbf{E}(x, y, z, k)$ let the amplitude functions in the integral representation (2.9) be denoted simply by $F_l(k_x, k_y, k)$ where l represents x, y, or z:

$$E_l(x, y, z, k) = \frac{1}{(2\pi)^2}\int_{-\infty}^{\infty}\int_{-\infty}^{\infty} F_l(k_x, k_y, k)\times$$
$$\times \exp[-i\{k_x x+k_y y+(k^2-k_x^{\,2}-k_y^{\,2})^{\frac{1}{2}}z\}]\, dk_x\, dk_y. \quad (2.11)$$

At the aperture plane $z = 0$ the two tangential components of electric field are then represented by

$$E_x(x, y, 0, k) = \frac{1}{(2\pi)^2}\int_{-\infty}^{\infty}\int_{-\infty}^{\infty} F_x(k_x, k_y, k)e^{-i(k_x x+k_y y)}\, dk_x\, dk_y, \quad (2.12a)$$

$$E_y(x, y, 0, k) = \frac{1}{(2\pi)^2}\int_{-\infty}^{\infty}\int_{-\infty}^{\infty} F_y(k_x, k_y, k)e^{-i(k_x x+k_y y)}\, dk_x\, dk_y. \quad (2.12b)$$

Within the aperture A this represents the source. It is in the form of a two-dimensional Fourier transform, the inverse of which determines two of the six unknown amplitude functions:

$$F_x(k_x, k_y, k) = \int_{-\infty}^{\infty}\int_{-\infty}^{\infty} E_x(x, y, 0, k)e^{i(k_x x+k_y y)}\, dx\, dy$$
$$= \iint_A E_x(x, y, 0, k)e^{i(k_x x+k_y y)}\, dx\, dy, \quad (2.13a)$$

$$F_y(k_x, k_y, k) = \int_{-\infty}^{\infty}\int_{-\infty}^{\infty} E_y(x, y, 0, k)e^{i(k_x x+k_y y)}\, dx\, dy$$
$$= \iint_A E_y(x, y, 0, k)e^{i(k_x x+k_y v)}\, dx\, dy. \quad (2.13b)$$

The range of integration in (2.13) has been limited to just the area of the aperture because of the fact that $E_x(x, y, 0, k)$ and $E_y(x, y, 0, k)$ must each vanish everywhere on the conducting sheet. Functions of the form (2.13) will be called *aperture-limited*.

The unknown amplitude function $F_z(k_x, k_y, k)$ can now be determined in terms of the two known amplitude functions $F_x(k_x, k_y, k)$ and $F_y(k_x, k_y, k)$ from the fact that $\nabla . \mathbf{E}$ is zero (obtained from (2.3*b*), since the divergence of the curl of any vector is zero). Calculating $\nabla . \mathbf{E}$ by differentiating the integral representation (2.11) for each of the three components and then setting it equal to zero,

$$\frac{-i}{(2\pi)^2}\int\limits_{-\infty}^{\infty}\int\limits_{-\infty}^{\infty}\{k_x F_x(k_x, k_y, k)+k_y F_y(k_x, k_y, k)+(k^2-k_x^2-k_y^2)^{\frac{1}{2}}F_z(k_x, k_y, k)\}\times$$
$$\times\exp[-i\{k_x x+k_y y+(k^2-k_x^2-k_y^2)^{\frac{1}{2}}z\}]\,dk_x\,dk_y = 0. \quad (2.14)$$

At $z = 0$ this integral is a two-dimensional Fourier transform of the bracketed expression. Hence the bracketed expression itself must be identically zero for all k_x and k_y, from which it follows that

$$F_z(k_x, k_y, k) = -\frac{k_x F_x(k_x, k_y, k)+k_y F_y(k_x, k_y, k)}{(k^2-k_x^2-k_y^2)^{\frac{1}{2}}}. \quad (2.15)$$

The three components of $\mathbf{H}(x, y, z, k)$ can now be obtained from (2.3*a*) by calculating $\nabla\times\mathbf{E}(x, y, z, k)$. All six of the rectangular components of the electric and magnetic field functions are then as follows:†

$$E_x(x, y, z, k) = \frac{1}{(2\pi)^2}\int\limits_{-\infty}^{\infty}\int\limits_{-\infty}^{\infty}F_x\times$$
$$\times\exp[-i\{k_x x+k_y y+(k^2-k_x^2-k_y^2)^{\frac{1}{2}}z\}]\,dk_x\,dk_y, \quad (2.16a)$$

$$E_y(x, y, z, k) = \frac{1}{(2\pi)^2}\int\limits_{-\infty}^{\infty}\int\limits_{-\infty}^{\infty}F_y\times$$
$$\times\exp[-i\{k_x x+k_y y+(k^2-k_x^2-k_y^2)^{\frac{1}{2}}z\}]\,dk_x\,dk_y, \quad (2.16b)$$

$$E_z(x, y, z, k) = -\frac{1}{(2\pi)^2}\int\limits_{-\infty}^{\infty}\int\limits_{-\infty}^{\infty}\frac{k_x F_x+k_y F_y}{(k^2-k_x^2-k_y^2)^{\frac{1}{2}}}\times$$
$$\times\exp[-i\{k_x x+k_y y+(k^2-k_x^2-k_y^2)^{\frac{1}{2}}z\}]\,dk_x\,dk_y, \quad (2.16c)$$

† Equivalent expressions have been obtained by various authors. See, for example: J. Brown. A theoretical analysis of some errors in aerial measurements. *Proc. Instn electr. Engrs*, **105C**, pp. 343–51, February 1958; G. Borgiotti. Fourier transform method in aperture antenna problems. *Alta Frequenza*, **32**, pp. 196*E*–204*E*, November 1963.

$$H_x(x, y, z, k) = -\frac{1}{(2\pi)^2 kZ_0}\int_{-\infty}^{\infty}\int_{-\infty}^{\infty}\frac{k_x k_y F_x + (k^2 - k_x^2)F_y}{(k^2 - k_x^2 - k_y^2)^{\frac{1}{2}}}\times$$
$$\times\exp[-i\{k_x x + k_y y + (k^2 - k_x^2 - k_y^2)^{\frac{1}{2}} z\}]\,dk_x\,dk_y, \quad (2.16d)$$

$$H_y(x, y, z, k) = \frac{1}{(2\pi)^2 kZ_0}\int_{-\infty}^{\infty}\int_{-\infty}^{\infty}\frac{(k^2 - k_y^2)F_x + k_x k_y F_y}{(k^2 - k_x^2 - k_y^2)^{\frac{1}{2}}}\times$$
$$\times\exp[-i\{k_x x + k_y y + (k^2 - k_x^2 - k_y^2)^{\frac{1}{2}} z\}]\,dk_x\,dk_y, \quad (2.16e)$$

$$H_z(x, y, z, k) = -\frac{1}{(2\pi)^2 kZ_0}\int_{-\infty}^{\infty}\int_{-\infty}^{\infty}(k_y F_x - k_x F_y)\times$$
$$\times\exp[-i\{k_x x + k_y y + (k^2 - k_x^2 - k_y^2)^{\frac{1}{2}} z\}]\,dk_x\,dk_y, \quad (2.16f)$$

where Z_0 denotes the characteristic impedance $(\mu_0/\epsilon_0)^{\frac{1}{2}}$ of free space.

Each of the six complex amplitude functions in (2.16) has been expressed in terms of just two independent scalar functions $F_x(k_x, k_y, k)$ and $F_y(k_x, k_y, k)$ that are determined, in turn, by the rectangular components $E_x(x, y, 0, k)$ and $E_y(x, y, 0, k)$, respectively, of the source. Thus, the field functions everywhere in the half-space excited by a planar source are determined exactly and uniquely by the source alone.

2.2. Interpretation of the elementary solutions

The integral representation (2.16) for the field functions of a radiating planar source was constructed as a straightforward solution of a boundary value problem by superposing all physically admissible elementary solutions of the scalar wave equation. No physical interpretation of the individual elementary solutions was given there because none was needed. But when the energy storage properties of planar antennas are determined in the next chapter it will be helpful to have a clear conceptual picture of the physical nature of those elementary solutions. Of particular importance is the distinction between the solutions that contribute to radiative power and those that contribute to reactive power. For that reason each of the two cases will be examined here in some detail.

The elementary solutions for all six components of the field functions have a common functional dependence upon the position coordinates (x, y, z), namely,

$$\exp[-i\{k_x x + k_y y + (k^2 - k_x^2 - k_y^2)^{\frac{1}{2}} z\}], \quad (2.17)$$

all differences between them residing solely in their complex amplitudes $F(k_x, k_y, k)$. It is the physical character of the position-dependent function (2.17) that is of principal interest here.

The distinctive differences between radiative and reactive solutions arise from an essential change that occurs in the character of the function (2.17) for real values of k_x, k_y, and k when the value of $(k^2-k_x^2-k_y^2)^{\frac{1}{2}}$ changes from a real to an imaginary number. Thus, these are the two cases to be examined.

(i) *Radiative solutions* $(k_x^2+k_y^2 < k^2)$

Inside the circle $k_x^2+k_y^2 = k^2$ the value of k_z, obtained from the square-root (2.7), is purely real. Hence the physical interpretation of (2.17), written as

$$e^{-i(k_x x+k_y y+k_z z)}, \tag{2.18}$$

is that of a wave (because it satisfies the wave equation) with constant magnitude (its absolute value is unity everywhere in space) and with a constant

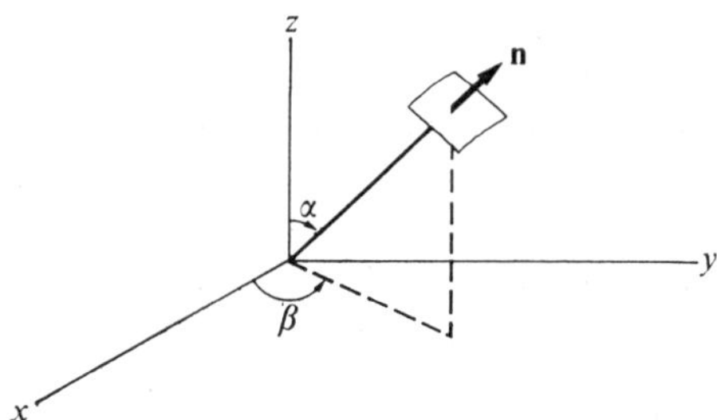

FIG. 2.2. Uniform plane wave with unit normal **n** in the direction (α, β).

phase front described by

$$k_x x+k_y y+k_z z = \text{constant}. \tag{2.19}$$

Since this is an equation for a plane, the function (2.18) must represent a uniform, unattenuated plane wave. Its direction of propagation is determined by the coefficients of the position variables x, y, and z, each of which must be proportional to a direction cosine n_x, n_y, and n_z, respectively:

$$k_x = Cn_x, \qquad k_y = Cn_y, \qquad k_z = Cn_z. \tag{2.20}$$

The direction cosines are the rectangular components of the unit vector **n** in the direction (α, β) normal to the plane of the phase front, as in Fig. 2.2, hence

$$n_x = \sin\alpha\cos\beta, \tag{2.21a}$$

$$n_y = \sin\alpha\sin\beta, \tag{2.21b}$$

$$n_z = \cos\alpha, \tag{2.21c}$$

and

$$n_x^2+n_y^2+n_z^2 = 1. \tag{2.22}$$

The constant C in (2.20) is then found from (2.6) to have the value k (the other root $-k$ corresponds to an incoming wave, which would violate the radiation condition), from which it follows that

$$k_x = k \sin \alpha \cos \beta, \tag{2.23a}$$

$$k_y = k \sin \alpha \sin \beta, \tag{2.23b}$$

$$k_z = k \cos \alpha. \tag{2.23c}$$

Thus, the constants k_x, k_y, and k_z that arose from separation of the wave equation are now seen to be interpretable physically as the rectangular components of a vector propagation constant,

$$\mathbf{k} = k\mathbf{n} = k_x\hat{\mathbf{x}}+k_y\hat{\mathbf{y}}+k_z\hat{\mathbf{z}}, \tag{2.24}$$

whose length is k and whose direction is that of the unit normal $\mathbf{n}$ to the plane phase front. The velocity of the phase front in that direction is the phase velocity

$$v_p = \frac{\omega}{(k_x^2+k_y^2+k_z^2)^{\frac{1}{2}}} = \frac{\omega}{k}, \tag{2.25}$$

which is simply the velocity c of light in free space.

(ii) *Reactive solutions* $(k_x^2+k_y^2 > k^2)$

Here the value of k_z obtained from (2.7) becomes purely imaginary. Its value, from (2.10), is

$$k_z = -i(k_x^2+k_y^2-k^2)^{\frac{1}{2}}. \tag{2.26}$$

The elementary solutions are then of the quite different form

$$\exp\{-(k_x^2+k_y^2-k^2)^{\frac{1}{2}}z\}e^{-i(k_x x+k_y y)}. \tag{2.27}$$

They no longer represent unattenuated plane waves but, instead, plane waves that decay exponentially with distance z from the aperture plane. The surfaces of constant magnitude are described by

$$(k_x^2+k_y^2-k^2)^{\frac{1}{2}}z = \text{constant}, \tag{2.28}$$

which are planes parallel to the aperture plane (Fig. 2.3*a*), whereas the surfaces of constant phase are described by

$$k_x x+k_y y = \text{constant}, \tag{2.29}$$

which are planes perpendicular to the planes of constant magnitude (Fig. 2.3*b*). Thus they represent nonuniform plane waves whose phase front always moves parallel to the aperture plane. The propagation constant of the phase front is $(k_x^2+k_y^2)^{\frac{1}{2}}$, which is always greater than the free space propagation constant k of uniform plane waves. The phase velocity of these

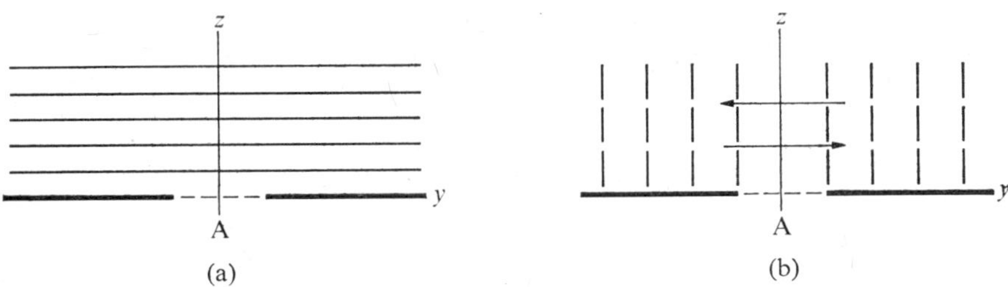

FIG. 2.3. Planes of (a) constant magnitude, and (b) constant phase, of a nonuniform (reactive) plane wave. The phase front propagates parallel to the aperture plane.

nonuniform plane waves is then

$$v_p = \frac{\omega}{(k_x^2+k_y^2)^{\frac{1}{2}}} < \frac{\omega}{k}, \tag{2.30}$$

which is always less than the velocity of light. For this reason they are frequently referred to as *slow waves*. And because of their exponential attenuation with distance from the aperture plane they are also frequently referred to as *evanescent waves*.

The plane wave nature of the elementary solutions of the wave equation has led to the descriptive name *angular spectrum* for the complex amplitude function $F(k_x, k_y, k)$. This terminology was used by Woodward and Lawson,† and again by Booker and Clemmow‡ in a paper that provided some interesting insight into the concept of angular spectrum. Some of the technical details have been elaborated upon more recently by Clemmow.§

2.3. Visible and invisible regions

At increasingly great distances from any planar source the contributions of the evanescent waves to the field functions become vanishingly small compared to those of the uniform plane waves. As a consequence only the uniform plane waves contribute to the radiation pattern; i.e. the evanescent waves do not radiate. Thus the circle $k_x^2+k_y^2 = k^2$ in the *wavenumber* plane of the real variables k_x and k_y separates the region $k_x^2+k_y^2 < k^2$ of uniform plane waves that are *visible* in the far field from the region $k_x^2+k_y^2 > k^2$ of nonuniform plane waves that are *invisible* in the far field, as indicated in Fig. 2.4.

Every point in the wavenumber plane has a unique one-to-one correspondence to a pair of points on the real axes of the complex k_x and k_y planes.

† P. M. Woodward and J. D. Lawson, *loc. cit.*

‡ H. G. Booker and P. C. Clemmow. The concept of an angular spectrum of plane waves, and its relation to that of polar diagram and aperture distribution. *Proc. Instn electr. Engrs*, **97**, Part III, pp. 11–17, January 1950.

§ P. C. Clemmow. *The plane wave spectrum representation of electromagnetic fields*, Pergamon Press, Oxford, 1966.

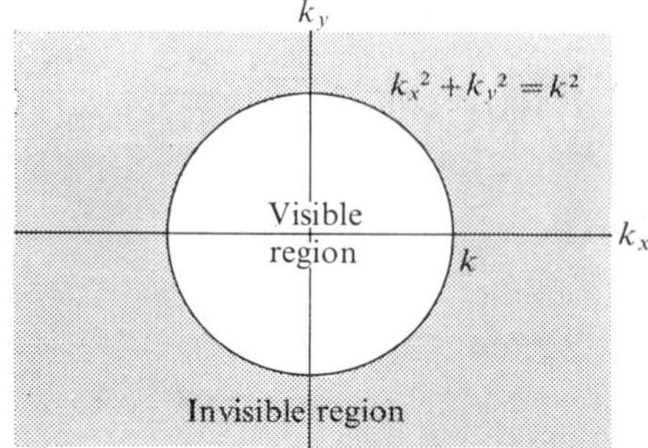

FIG. 2.4. Visible and invisible regions of the (k_x, k_y) plane.

And each of these pairs of points can be put into a unique one-to-one correspondence with a pair of points located on certain contours in the complex α and β planes. The complex angles α and β can be expressed in terms of k_x and k_y by solving (2.23*a*) and (2.23*b*) simultaneously:

$$\alpha = \text{arc sin}\{\pm(k_x^2+k_y^2)^{\frac{1}{2}}/k\}, \tag{2.31a}$$

$$\beta = \text{arc tan}(k_y/k_x). \tag{2.31b}$$

To make the correspondence unique let the α- and β-plane contours be chosen such that

$$-\infty < k \sin \alpha < \infty, \tag{2.32a}$$

$$-\pi/2 \leq \beta \leq \pi/2. \tag{2.32b}$$

Then the plane-polar coordinates of any point in the wavenumber plane will be $(k \sin \alpha, \beta)$. The point at the origin corresponds physically to a uniform plane wave moving in the broadside direction $\alpha = 0$. And the circle $k_x^2+k_y^2 = k^2$ corresponds to a continuum of uniform plane waves moving along the aperture plane at velocity c, each wave moving in an azimuthal direction β at the real polar angle $\alpha = +\pi/2$ in the right half-plane and $\alpha = -\pi/2$ in the left half-plane. All points outside of the circle (the invisible region) correspond to evanescent plane waves moving along the aperture plane in an azimuthal direction β at velocities less than c. Here the polar angle α is complex, its real part $\text{Re}\,\alpha = \pm\pi/2$ indicating the polar direction in which the phase front moves. Its imaginary part is found to be

$$\text{Im}\,\alpha = \pm\text{arc cosh}\{(k_x^2+k_y^2)^{\frac{1}{2}}/k\} \quad \text{at} \quad \text{Re}\,\alpha = \pm\pi/2, \tag{2.33}$$

after the sign ambiguity arising from the arc cosh function has been resolved by the tedious but straightforward process of imposing the still unused condition (2.23*c*).

Thus there are three different, but exactly equivalent, graphical representations for any point in the wavenumber plane. One is just the point (k_x, k_y) itself in the wavenumber plane, as indicated by Fig. 2.5a. Another is a pair of points on the real axes of the complex k_x and k_y planes, as in Fig. 2.5b. And the third is a pair of points on the complex α- and β-plane contours shown in Fig. 2.5c. Any point on the α-plane contour corresponds to a semicircle of points at radius $k \sin \alpha$ in the wavenumber plane. The real interval $-\pi/2 < \alpha < \pi/2$ thus defines the visible region of radiating plane waves,

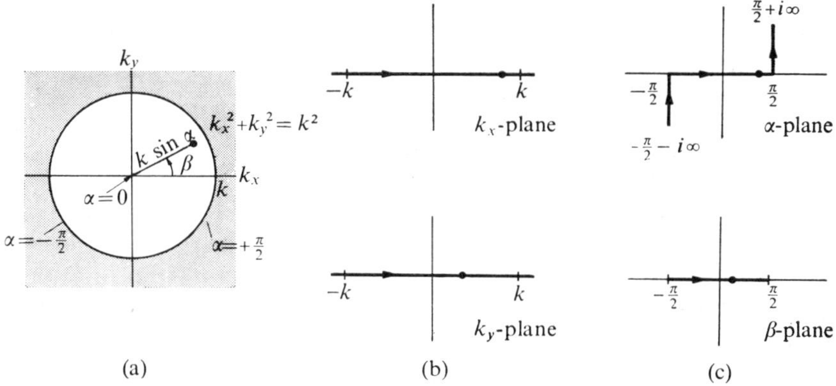

FIG. 2.5. Equivalent representations for any point in the wavenumber plane: (a) a single point in the (k_x, k_y) plane; (b) a pair of points on the real axes of the complex k_x and k_y planes; (c) a pair of points on the contours shown in the complex α and β planes.

each radiating wave corresponding to a point inside the circle of radius k in the wavenumber plane. The rest of the contour running from $-\pi/2-i\infty$ to $\pi/2+i\infty$ defines the invisible region of nonradiating (evanescent) plane waves that hug the aperture plane. Each of these nonradiating waves corresponds to a point outside the circle of radius k in the wavenumber plane. The endpoints $-\pi/2-i\infty$ and $\pi/2+i\infty$ of the contour correspond to all points on the circle of infinite radius. They represent the limiting case of evanescent plane waves with infinite attenuation, from (2.27), moving at zero phase velocity, from (2.30). In contrast to the complex polar angle α the azimuthal angle β is always real and always represents a true physical direction in space, for evanescent as well as for uniform plane waves. And it has a particularly simple connection with any point in the wavenumber plane in that it represents directly the angle of the polar coordinates of that point, whereas α determines the radius of that point only indirectly through the function $k \sin \alpha$.

Any point in the visible region of the wavenumber plane can be pictured as the two-dimensional projection of the tip of the real three-dimensional vector propagation constant **k**, as will be illustrated shortly in Fig. 2.6b. This geometrical picture breaks down for points in the invisible region, however, because the component k_z, and hence the vector **k** itself, becomes complex.

2.4. Pattern space factor[9]

The elementary solutions have been found to be interpretable physically as radiating plane waves in the visible region and as nonradiating plane waves in the invisible region. Now it will be shown that the complex amplitude $F(k_x, k_y, k)$ of those plane waves is interpretable over the visible region in terms of the radiation pattern. This relationship between the radiation pattern and the plane wave amplitude is essential to source synthesis because it provides the analytic connection needed between the radiation pattern and all of the other properties of the source.

By the radiation pattern of an antenna is meant the angular behaviour of the field functions as the distance r from the source approaches infinity. It is a vector function, since the field functions are vectors. From (2.16) the vector electric field function $\mathbf{E}(x, y, z, k)$, for example, is seen to be

$$\mathbf{E}(x, y, z, k) = \frac{1}{(2\pi)^2}\int_{-\infty}^{\infty}\int_{-\infty}^{\infty} \mathbf{F}(k_x, k_y, k)\times$$

$$\times \exp[-i\{k_x x + k_y y + (k^2 - k_x^2 - k_y^2)^{\frac{1}{2}} z\}]\, dk_x\, dk_y, \quad (2.34)$$

where

$$\mathbf{F}(k_x, k_y, k) = \hat{\mathbf{x}} F_x(k_x, k_y, k) + \hat{\mathbf{y}} F_y(k_x, k_y, k) - \hat{\mathbf{z}} \frac{k_x F_x(k_x, k_y, k) + k_y F_y(k_x, k_y, k)}{(k^2 - k_x^2 - k_y^2)^{\frac{1}{2}}} \quad (2.35)$$

This vector relationship (2.34) has meaning only when the two vector functions $\mathbf{E}(x, y, z, k)$ and $\mathbf{F}(k_x, k_y, k)$ are resolved into rectangular components, as they were in (2.16), since only in rectangular coordinates can the unit vectors in two independent vector spaces be assigned a one-to-one correspondence. Any correspondence is admissible, the simplest one being to assign the unit vector $\hat{\mathbf{k}}_x$ to $\hat{\mathbf{x}}$, $\hat{\mathbf{k}}_y$ to $\hat{\mathbf{y}}$, and $\hat{\mathbf{k}}_z$ to $\hat{\mathbf{z}}$, as suggested by Fig. 2.6. Then the component $F_{k_x}(k_x, k_y, k)$ must be assigned to $E_x(x, y, z, k)$, $F_{k_y}(k_x, k_y, k)$ to $E_y(x, y, z, k)$, and $F_{k_z}(k_x, k_y, k)$ to $E_z(x, y, z, k)$. The functions $F_{k_x}(k_x, k_y, k)$, $F_{k_y}(k_x, k_y, k)$, and $F_{k_z}(k_x, k_y, k)$ have been and will continue to be denoted by $F_x(k_x, k_y, k)$, $F_y(k_x, k_y, k)$, and $F_z(k_x, k_y, k)$, respectively, for simplicity of notation. This particular assignment of the unit vectors has the advantage that it leads to the usual connection $k_x = k \sin\theta \cos\phi$ and $k_y = k \sin\theta \sin\phi$, from Fig. 2.6, between a point (k_x, k_y) in the wavenumber plane and the direction (θ, ϕ) to a far-field point in space.

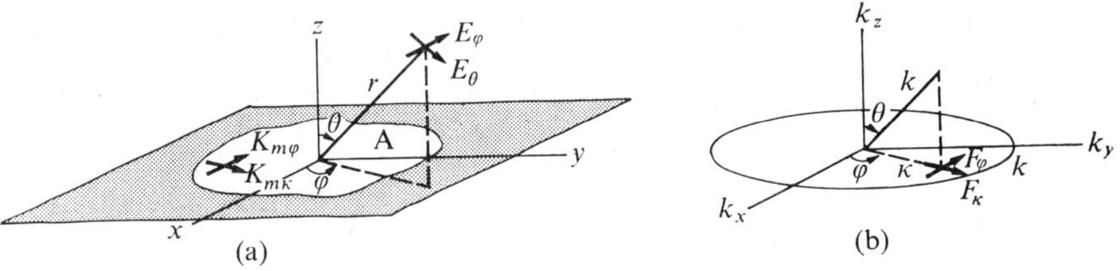

FIG. 2.6. (a) Far-field spherical components $E_\theta(r, \theta, \phi, k)$ and $E_\phi(r, \theta, \phi, k)$ produced by an arbitrary source over the aperture A. (b) The corresponding polar components $F_\kappa(\kappa, \phi, k)$ and $F_\phi(\kappa, \phi, k)$ of the pattern space factor in the visible region $\kappa \leqslant k$ of the (k_x, k_y) plane.

At increasingly great distances r each of the three rectangular components of $\mathbf{E}(x, y, z, k)$ is shown in Appendix 1 to approach asymptotically the value

$$E_l(x, y, z, k) \sim \frac{ik}{2\pi}\frac{e^{-ikr}}{r}\cos\theta F_l(k\sin\theta\cos\phi, k\sin\theta\sin\phi, k) \tag{2.36}$$

in any given direction (θ, ϕ), where l denotes x, y, or z. From this simple asymptotic relationship between the rectangular components of $\mathbf{E}(x, y, z, k)$ and $\mathbf{F}(k_x, k_y, k)$ an equally simple relationship will now be obtained between the spherical angular components of $\mathbf{E}(x, y, z, k)$ and the plane-polar components of $\mathbf{F}(k_x, k_y, k)$.

First, represent the spherical-angular components of $\mathbf{E}(x, y, z, k)$ in terms of rectangular components, as in Fig. 2.6a,

$$E_\theta(r, \theta, \phi, k) = E_x(x, y, z, k)\cos\theta\cos\phi + \\ + E_y(x, y, z, k)\cos\theta\sin\phi - E_z(x, y, z, k)\sin\theta, \tag{2.37a}$$

$$E_\phi(r, \theta, \phi, k) = -E_x(x, y, z, k)\sin\phi + E_y(x, y, z, k)\cos\phi, \tag{2.37b}$$

and then represent the rectangular components of $\mathbf{F}(k_x, k_y, k)$ in the asymptotic expression (2.36) in terms of plane-polar components, as in Fig. 2.6b,

$$F_x(k_x, k_y, k) = F_\kappa(\kappa, \phi, k)\cos\phi - F_\phi(\kappa, \phi, k)\sin\phi, \tag{2.38a}$$

$$F_y(k_x, k_y, k) = F_\kappa(\kappa, \phi, k)\sin\phi + F_\phi(\kappa, \phi, k)\cos\phi, \tag{2.38b}$$

where κ denotes the radial variable $k\sin\theta$ in the wavenumber plane. When these are combined one obtains the remarkably simple asymptotic relationship

$$E_\theta(r, \theta, \phi, k) \sim \frac{ik}{2\pi}\frac{e^{-ikr}}{r}F_\kappa(\kappa, \phi, k), \tag{2.39a}$$

$$E_\phi(r, \theta, \phi, k) \sim \frac{ik}{2\pi}\frac{e^{-ikr}}{r}\cos\theta F_\phi(\kappa, \phi, k). \tag{2.39b}$$

The radial factor e^{-ikr}/r verifies that the far-field is in the form of an outgoing wave (obtained by the choice of negative signs in the exponents of (2.8)) of decreasing magnitude, as required by the radiation condition.

The radiation pattern for the spherical component $E_\theta(r, \theta, \phi, k)$ is thus just $F_\kappa(\kappa, \phi, k)$ alone, while that for $E_\phi(r, \theta, \phi, k)$ is just the product $\cos\theta F_\phi(\kappa, \phi, k)$. Their interpretation is as follows. From the vector relationship obtained from (2.13) it is clear that each of the plane-wave amplitude components $F_\kappa(\kappa, \phi, k)$ and $F_\phi(\kappa, \phi, k)$ is determined solely by the corresponding source component $E_\kappa(x, y, 0, k)$ and $E_\phi(x, y, 0, k)$, respectively. And each of these source components, in turn, can be thought of as terminating on a sheet of equivalent magnetic surface current density $-K_{m\phi}(x, y, k)$ and $K_{m\kappa}(x, y, k)$, respectively. Hence, at a great distance from the source the electric field component $E_\theta(r, \theta, \phi, k)$ in (2.39*a*) can be thought of as being produced by $-K_{m\phi}(x, y, k)$ and the other electric field component $E_\phi(r, \theta, \phi, k)$ in (2.39*b*) as being produced by $K_{m\kappa}(x, y, k)$. For an infinitesimally small element of the magnetic current $K_{m\phi}(x, y, k)$ the far-field component $E_\theta(r, \theta, \phi, k)$ is independent of θ, whereas for an infinitesimally small element of the magnetic current $K_{m\kappa}(x, y, k)$ the far-field component $E_\phi(r, \theta, \phi, k)$ varies as $\cos\theta$. Together these two far-field components of an infinitesimally small element (a Hertzian dipole) represent the *element factor* of the radiation pattern. They are well known properties of the radiation pattern of elementary Hertzian dipoles, the latter of which will be shown in section 2.7 to come directly from the field solution obtained in section 2.1. It is evident, then, that the function $F_\kappa(\kappa, \phi, k)$ in (2.39*a*) accounts for the sole effect of the spatial distribution of the source upon the far-field component $E_\theta(r, \theta, \phi, k)$, while the function $F_\phi(\kappa, \phi, k)$ in (2.39*b*) accounts for the sole effect of the spatial distribution of the source upon the far-field component $E_\phi(r, \theta, \phi, k)$. From this it is concluded that the plane-polar components $F_\kappa(\kappa, \theta, \phi)$ and $F_\phi(\kappa, \theta, \phi)$ of the plane-wave amplitude function represent the *pattern space factor* for the radiation pattern of the spherical components $E_\theta(r, \theta, \phi, k)$ and $E_\phi(r, \theta, \phi, k)$, respectively, of the far field.

It is important to recognize that this resolution of the source into polar components $E_\kappa(x, y, 0, k)$ and $E_\phi(x, y, 0, k)$ in the wavenumber plane (k_x, k_y) causes those components to vary with the direction (θ, ϕ) of the far-field point of observation. This is quite unusual, and is totally different from its customary resolution into components in local coordinates. But it has the distinct advantage of giving the most direct physical interpretation of the functions $F_\kappa(\kappa, \phi, k)$ and $F_\phi(\kappa, \phi, k)$ in terms of the source itself.

In summary, the two spherical components of the radiation pattern of any planar source are each seen to separate naturally into a product of two factors. One is the element factor that can be interpreted simply as the radiation pattern of an infinitesimal element of equivalent magnetic current. It accounts for the polarization properties of the source, and it always has the value unity for the θ-component and $\cos\theta$ for the ϕ-component. The other is the pattern space factor that accounts for the spatial properties of the source. The pattern space factor provides the only controllable part of the radiation pattern. Hence the pattern space factor is the focal point of the antenna synthesis problem.

2.5. Complementary antennas

Suppose that the conducting planar sheet containing the aperture is further idealized to be of zero thickness and that the source is free to radiate equally on both sides. Under these conditions, and only under these conditions, an extremely simple relationship exists between the field functions of

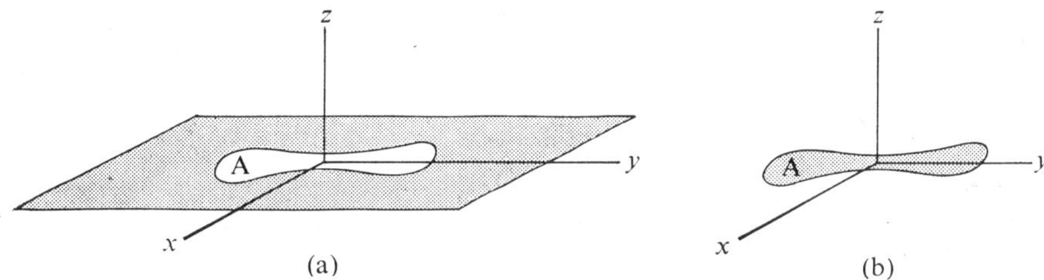

FIG. 2.7. (a) An arbitrary planar aperture A, and (b) its complementary aperture.

the idealized slot antenna and those of the idealized complementary dipole antenna that would be obtained by interchanging the conducting and non-conducting portions of the sheet. That relationship will be derived here. Its importance lies in the fact that it provides a simple connection between all properties of complementary idealized slot and dipole antennas. One of these properties is Booker's impedance relation that will be derived in the next section.

If an aperture A were cut in a conducting sheet of zero thickness, as in Fig. 2.7a, the conducting part removed would be the *complementary aperture* A shown in Fig. 2.7b. The conducting sheet containing the aperture becomes an idealized slot antenna when excited by a transmission line connected to the two infinitesimally close sides at $y = 0$, and the complementary aperture becomes an idealized dipole antenna when cut into two halves at $y = 0$ and excited by a transmission line connected to the two infinitesimally close sides of the gap. In either case the field functions on both sides of the aperture plane $z = 0$ will again be those solutions of the system equations (2.3) that satisfy the boundary condition at $z = 0$ and the radiation condition at infinity. The boundary condition used in section 2.1 to obtain the field functions of a planar source was a condition on tangential E rather than H, but that was simply a matter of convenience. Tangential E was chosen there because of the simplicity that results from the fact that it must vanish on the plane $z = 0$ outside the aperture A. Exactly the same simplicity would result in the case of the complementary aperture if, instead, tangential H were to vanish outside of the complementary aperture A, since the field functions everywhere are fully determined by either tangential E or H over the entire plane. It will now be shown that tangential H does, in fact, vanish everywhere in the hole (i.e. in the nonconducting part of the aperture plane) if the conducting sheet is of zero thickness and is free to radiate equally on both sides. But if these two conditions are not satisfied then tangential H need not vanish in the hole and the field functions need not be determined by tangential H over just the conducting sheet alone. For example, it need not apply even to the practical case of a slot or a dipole antenna of nonzero thickness, nor to that of an aperture enclosed on one side by a conducting cavity.

The fact that tangential H will be precisely zero everywhere in the hole, for apertures as well as for complementary apertures, will be established by a method used by Bethe.† Let the field functions above the aperture plane ($z > 0$) be designated by $\mathbf{E}(x, y, z, k)$ and $\mathbf{H}(x, y, z, k)$ and those below ($z < 0$) by $\mathbf{E}'(x, y, z, k)$ and $\mathbf{H}'(x, y, z, k)$. They must satisfy the system equations (2.3) and the boundary conditions that E_{tan} and E'_{tan} both vanish on the conducting part of the aperture plane $z = 0$ and that all of the field components are continuous through the hole. Suppose, now, that the field $\mathbf{E}$, $\mathbf{H}$ satisfies the system equations and that E_{tan} is zero on the conductor. Then it is easily verified that all requirements will be satisfied if the field $\mathbf{E}'$, $\mathbf{H}'$ is such that

$$E_x(x, y, z, k) = E_x'(x, y, -z, k), \tag{2.40a}$$

$$E_y(x, y, z, k) = E_y'(x, y, -z, k), \tag{2.40b}$$

$$E_z(x, y, z, k) = -E_z'(x, y, -z, k), \tag{2.40c}$$

$$H_x(x, y, z, k) = -H_x'(x, y, -z, k), \tag{2.40d}$$

$$H_y(x, y, z, k) = -H_y'(x, y, -z, k), \tag{2.40e}$$

$$H_z(x, y, z, k) = H_z'(x, y, -z, k). \tag{2.40f}$$

At $z = 0$,

$$H_x(x, y, 0, k) = -H_x'(x, y, 0, k), \tag{2.41a}$$

$$H_y(x, y, 0, k) = -H_y'(x, y, 0, k). \tag{2.41b}$$

But the tangential components of magnetic field must be continuous at the hole:

$$H_x(x, y, 0, k) = H_x'(x, y, 0, k), \tag{2.42a}$$

$$H_y(x, y, 0, k) = H_y'(x, y, 0, k). \tag{2.42b}$$

Hence it is concluded that

$$H_x(x, y, 0, k) = H_y(x, y, 0, k) = 0 \quad \text{in the hole.} \tag{2.43}$$

† H. A. Bethe. Theory of diffraction by small holes. *Phys. Rev.* **66**, pp. 163–82, October 1944.

Since the argument was based upon the assumption of symmetry of the field functions about the plane $z = 0$ it is clear that (2.43) need not hold when that symmetry is absent.

A curious consequence of this result is that, in principle, idealized slot or dipole antennas do not radiate through their apertures. This is evident from the fact that for such antennas

$$E_{\text{tan}}|_{\text{conductor}} = H_{\text{tan}}|_{\text{hole}} = 0, \tag{2.44}$$

from which it follows that

$$-\tfrac{1}{2} \iint\limits_{\substack{\text{aperture} \\ \text{plane}}} \mathbf{E}\times\mathbf{H}^*.\mathbf{n}\, da = 0 \tag{2.45}$$

for either type of idealized antenna; i.e. the complex power flow through the aperture must be zero. Thus there must be an essential difference in the physical mechanism of radiation between idealized slot or dipole antennas, on the one hand, and cavity-excited aperture-antennas on the other. Cavity-excited aperture-antennas can have an actual flow of power through the aperture, since there is no symmetry about the aperture plane to force H_{tan} to be zero in the aperture. But no power can flow through the aperture of an idealized slot or dipole antenna, so it appears that the role of the aperture in that case is just to guide power from the probe that excites the aperture. If this is true, all power from an idealized slot or dipole antenna must originate from just the exciting probe rather than from the aperture as a whole. This conjecture is consistent with some calculations made by Schelkunoff and Friis† on perfectly conducting biconical antennas, from which they observed that the flow lines of time-average power all originate from the gap at the apex, even for the case of idealized infinitely thin antennas in which the lines of power flow give the impression of emerging from the conducting arms of the antenna itself.

The relationship between the field functions of an idealized slot antenna and those of its complementary dipole antenna is now easy to establish. Let the fields of the idealized slot antenna be denoted by $\mathbf{E}^{\text{s}}(x, y, z, k)$, $\mathbf{H}^{\text{s}}(x, y, z, k)$ and let the fields of the complementary dipole antenna be denoted by $\mathbf{E}^{\text{d}}(x, y, z, k)$, $\mathbf{H}^{\text{d}}(x, y, z, k)$. Each pair of fields can be determined uniquely from just the system equations (2.3) and the boundary conditions. Consider first the boundary conditions. For the slot antenna they require only that tangential E be zero in the plane outside of the aperture A and that the exciting voltage V^{s} at the input terminals have the value prescribed. And for the complementary dipole antenna they require only that tangential H be zero in the plane outside of the complementary aperture A

† S. A. Schelkunoff and H. T. Friis, *loc. cit.* pp. 124–5.

and that the exciting current I^d at the input terminals have the value prescribed for it. Except for a choice of scale factor that is determined by the excitation, then, the boundary conditions on $\mathbf{E}^s$ are identical to the boundary conditions on $\mathbf{H}^d$. Consider next the system equations (2.3). A simple calculation shows that they are invariant under the transformation

$$\mathbf{E} \to Z_0\mathbf{H}, \qquad Z_0\mathbf{H} \to -\mathbf{E}. \tag{2.46}$$

Thus it follows that the boundary-value problem for the idealized slot antenna is mathematically identical to that for the complementary dipole antenna. By properly choosing the relationship between the scale factors—i.e. the relationship between the exciting voltage V^s of the slot antenna and the exciting current I^d of the complementary dipole antenna—the two pairs of fields will be related everywhere in (x, y, z, k) by

$$\mathbf{E}^s(x, y, z, k) \equiv Z_0\mathbf{H}^d(x, y, z, k), \tag{2.47a}$$

$$Z_0\mathbf{H}^s(x, y, z, k) \equiv -\mathbf{E}^d(x, y, z, k). \tag{2.47b}$$

The required relationship between V^s and I^d will be seen later in (2.51).

The field functions for the complementary dipole can now be represented as a sum of plane waves in precisely the same way as in section 2.1 for the slot. Since the tangential components of $\mathbf{H}$, rather than of $\mathbf{E}$, must vanish in the plane outside of the complementary aperture A it is evident that the aperture distribution for the dipole must be proportional to $H_{\tan}$ rather than to $E_{\tan}$. By including the constant factor Z_0 the aperture distribution for the dipole must be identical to that for the idealized slot antenna, from (2.47a) evaluated at the aperture plane $z = 0$,

$$\left.\begin{aligned} E_x^s(x, y, 0, k) &\equiv Z_0H_x^d(x, y, 0, k) \quad (2.48a)\\ E_y^s(x, y, 0, k) &\equiv Z_0H_y^d(x, y, 0, k) \quad (2.48b)\end{aligned}\right\} \text{on } A.$$

Hence, if the two independent plane-wave amplitude functions $F_x^d(k_x, k_y, k)$ and $F_y^d(k_x, k_y, k)$ for the complementary dipole antenna are chosen to be related to the aperture distribution by

$$\begin{aligned} F_x^d(k_x, k_y, k) &= \int_{-\infty}^{\infty}\int_{-\infty}^{\infty} Z_0H_x^d(x, y, 0, k)e^{i(k_xx+k_yy)}\,dx\,dy \\ &= \iint_A Z_0H_x^d(x, y, 0, k)e^{i(k_xx+k_yy)}\,dx\,dy, \end{aligned} \tag{2.49a}$$

$$\begin{aligned} F_y^d(k_x, k_y, k) &= \int_{-\infty}^{\infty}\int_{-\infty}^{\infty} Z_0H_y^d(x, y, 0, k)e^{i(k_xx+k_yy)}\,dx\,dy \\ &= \iint_A Z_0H_y^d(x, y, 0, k)e^{i(k_xx+k_yy)}\,dx\,dy, \end{aligned} \tag{2.49b}$$

by analogy to (2.13) for slot antennas, it is evident that the amplitude functions for idealized slot and dipole antennas will be identical to each other:

$$F_x^{\mathrm{s}}(k_x, k_y, k) \equiv F_x^{\mathrm{d}}(k_x, k_y, k), \tag{2.50a}$$

$$F_y^{\mathrm{s}}(k_x, k_y, k) \equiv F_y^{\mathrm{d}}(k_x, k_y, k). \tag{2.50b}$$

Thus, all six components of the field functions for the complementary dipole antenna can be expressed as a sum of plane waves in precisely the same form as (2.16), after making the exchange (2.47) between the field functions of idealized slot and dipole antennas.

2.6. Booker's impedance relation

An important consequence of the connection (2.47) between the field functions of idealized slot and dipole antennas is the existence of a simple relationship between their input impedances. The essential features of a proof of that relationship of Booker's† are as follows.

An idealized slot antenna and its complementary dipole are illustrated in Fig. 2.8, including an infinitesimally short excitation probe and its transmission line (a hypothetical coaxial cable of zero thickness buried in the conducting sheet). The field functions $\mathbf{E}^{\mathrm{s}}$, $\mathbf{H}^{\mathrm{s}}$ of the slot and $\mathbf{E}^{\mathrm{d}}$, $\mathbf{H}^{\mathrm{d}}$ of the dipole that are produced by the excitation probes are assumed to be related as in (2.47). Then the corresponding relationships between the system functions for voltage and current at the two excitation probes are obtained by integrating (2.47*a*) from point 1 to point 2 on the $z > 0$ side,

$$V^{\mathrm{s}} = \int_1^2 \mathbf{E}^{\mathrm{s}}.\,d\mathbf{l} = \int_1^2 Z_0\mathbf{H}^{\mathrm{d}}.\,d\mathbf{l} = \tfrac{1}{2}Z_0 \oint \mathbf{H}^{\mathrm{d}}.\,d\mathbf{l} = \tfrac{1}{2}Z_0 I^{\mathrm{d}}, \tag{2.51}$$

and integrating (2.47*b*) along an infinitesimal half-circle C on the $z > 0$ side,

$$V^{\mathrm{d}} = \int_C \mathbf{E}^{\mathrm{d}}.\,d\mathbf{l} = \int_C (-Z_0\mathbf{H}^{\mathrm{s}}).\,d\mathbf{l} = -\tfrac{1}{2}Z_0 \oint \mathbf{H}^{\mathrm{s}}.\,d\mathbf{l} = \tfrac{1}{2}Z_0 I^{\mathrm{s}}. \tag{2.52}$$

The input impedance Z_{d} of the dipole is then

$$Z_{\mathrm{d}} = \frac{V^{\mathrm{d}}}{I^{\mathrm{d}}} = \frac{\dfrac{Z_0}{2}I^{\mathrm{s}}}{\dfrac{2}{Z_0}V^{\mathrm{s}}} = \frac{(\frac{1}{2}Z_0)^2}{Z_{\mathrm{s}}} \tag{2.53}$$

in terms of the input impedance Z_{s} of its complementary slot, from which one obtains *Booker's impedance relation*:

$$Z_{\mathrm{d}}Z_{\mathrm{s}} = (\tfrac{1}{2}Z_0)^2. \tag{2.54}$$

† H. G. Booker. Slot aerials and their relation to complementary wire aerials (Babinet's principle). *J. Instn elec. Engrs.*, Part IIIA, **93**, pp. 620–6, No. 4, 1946.

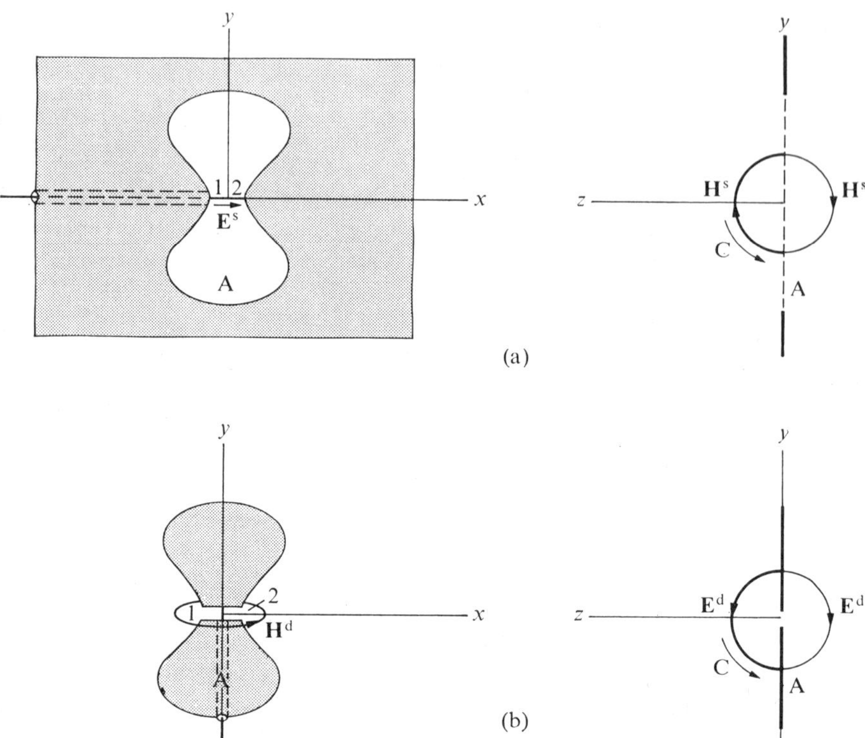

FIG. 2.8. (a) An idealized slot antenna, and (b) its complementary dipole.

Thus, the complex input impedance for either an idealized slot or dipole antenna of arbitrary size and shape is related simply to the complex input impedance of its complementary configuration by (2.54).

Booker's impedance relation is exact, but only for idealized slot and dipole antennas of zero thickness that radiate equally on both sides of the aperture. These restrictions arise because of the inherent dependence of (2.47) upon symmetry of the antenna configuration about the plane of the aperture. It might reasonably be expected to be approximately true if the thickness of the conducting sheet is sufficiently small, even though nonzero, since a certain degree of symmetry might still be preserved. But there is no reason at all for it to be even approximately true for a cavity-excited antenna that radiates into only a half space (e.g. a waveguide-fed slot in a conducting planar sheet) because such an antenna would be completely lacking in symmetry.

This relationship between the input impedances of complementary planar antennas led to the earliest recognition that antennas could be designed whose impedance is essentially independent of frequency over a very wide band. It appears that Mushiake† was the first to point out that any self-complementary antenna (a planar antenna that is its own complement) should, in principle, have a constant impedance of $\frac{1}{2}Z_0$ (equal to 60 π, or about 189 ohms) at all frequencies, because of Booker's impedance relation. There is an infinite variety of such antenna configurations, giving a constant radiation pattern as well, but all suffer from the practical disadvantage that their size must extend to infinity. The only practical one, in the sense that it retains all of its broadband properties when truncated to a finite size, was shown by Rumsey† to be the equiangular spiral.

2.7. Hertzian dipole

In the course of interpreting the radiation pattern in the far-field expression (2.39) as being the product of a pattern space factor by the radiation pattern

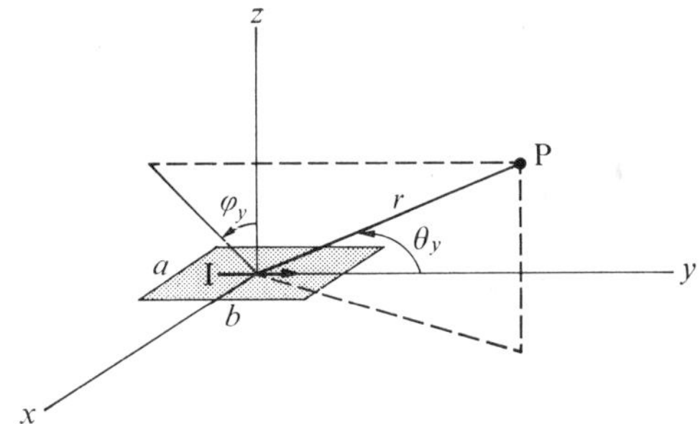

FIG. 2.9. Coordinate system for a planar Hertzian dipole of current I spread over a sheet of dimensions a and b.

of an elementary Hertzian dipole it was stated that the radiation pattern for a Hertzian dipole can be obtained directly from the field solution for planar sources that was developed in section 2.1. This will now be shown.

The Hertzian dipole will be represented as the limiting case of a y-directed electric current I spread uniformly over the top and bottom of a rectangular conducting sheet of dimensions a and b (Fig. 2.9) as both dimensions approach zero. Hence the aperture distribution has just the constant value

$$Z_0 H_x{}^{\mathrm{d}}(x, y, 0, k) = \frac{Z_0 I}{2a} \tag{2.55}$$

inside the aperture $-a/2 < x < a/2$, $-b/2 < y < b/2$, and zero outside. From (2.49) its pattern space factor is found to have but a single component,

$$F_x{}^{\mathrm{d}}(k_x, k_y, k) = \tfrac{1}{2}Z_0 I b \frac{\sin \frac{1}{2}k_x a}{\frac{1}{2}k_x a} \frac{\sin \frac{1}{2}k_y b}{\frac{1}{2}k_y b}. \tag{2.56}$$

† V. H. Rumsey. Frequency independent antennas, IRE International Convention Record, Part I, pp. 114–8, 1957. Also, *Frequency independent antennas*, Academic Press, New York, 1966; p. 17.

When the dimensions a and b become vanishingly small the pattern space factor becomes just the dipole moment Ib times half of the impedance of free space,

$$F_x^{\mathrm{d}}(k_x, k_y, k) = \tfrac{1}{2}Z_0 Ib, \tag{2.57}$$

independently of the value of k_x and k_y.

From the fields (2.16) and the correspondence (2.47) between the fields of idealized slots and dipoles the three rectangular components of magnetic field for a Hertzian dipole are seen to be

$$H_x^{\mathrm{d}}(x, y, z, k) = \frac{Ib}{2(2\pi)^2}\int_{-\infty}^{\infty}\int_{-\infty}^{\infty} \times$$

$$\times \exp[-i\{k_x x+k_y y+(k^2-k_x^{\,2}-k_y^{\,2})^{\frac{1}{2}}z\}]\, dk_x\, dk_y, \tag{2.58a}$$

$$H_y^{\mathrm{d}}(x, y, z, k) = 0, \tag{2.58b}$$

$$H_z^{\mathrm{d}}(x, y, z, k) = -\frac{Ib}{2(2\pi)^2}\int_{-\infty}^{\infty}\int_{-\infty}^{\infty}\frac{k_x}{(k^2-k_x^{\,2}-k_y^{\,2})^{\frac{1}{2}}} \times$$

$$\times \exp[-i\{k_x x+k_y y+(k^2-k_x^{\,2}-k_y^{\,2})^{\frac{1}{2}}z\}]\, dk_x\, dk_y. \tag{2.58c}$$

Each of the two non-zero components can be written as a partial derivative of a common double integral,

$$H_x^{\mathrm{d}}(x, y, z, k) = i\frac{Ib}{2(2\pi)^2}\frac{\partial}{\partial z} \times$$

$$\times \left\{\int_{-\infty}^{\infty}\int_{-\infty}^{\infty}\frac{\exp\{-i(k^2-k_x^{\,2}-k_y^{\,2})^{\frac{1}{2}}z\}}{(k^2-k_x^{\,2}-k_y^{\,2})^{\frac{1}{2}}}e^{-i(k_x x+k_y y)}\, dk_x\, dk_y\right\}, \tag{2.59a}$$

$$H_z^{\mathrm{d}}(x, y, z, k) = -i\frac{Ib}{2(2\pi)^2}\frac{\partial}{\partial x} \times$$

$$\times \left\{\int_{-\infty}^{\infty}\int_{-\infty}^{\infty}\frac{\exp\{-i(k^2-k_x^{\,2}-k_y^{\,2})^{\frac{1}{2}}z\}}{(k^2-k_x^{\,2}-k_y^{\,2})^{\frac{1}{2}}}e^{-i(k_x x+k_y y)}\, dk_x\, dk_y\right\}. \tag{2.59b}$$

By changing the variables of integration from rectangular to polar coordinates the double integral can be reduced to a Hankel transform of order zero whose value at any distance $r=(x^2+y^2+z^2)^{\frac{1}{2}}$ is known to be†

$$\int_{-\infty}^{\infty}\int_{-\infty}^{\infty}\frac{\exp\{-i(k^2-k_x^{\,2}-k_y^{\,2})^{\frac{1}{2}}z\}}{(k^2-k_x^{\,2}-k_y^{\,2})^{\frac{1}{2}}}e^{-i(k_x x+k_y y)}\,dk_x\,dk_y$$

$$=2\pi\int_0^{\infty}\frac{\exp\{-i(k^2-\kappa^2)^{\frac{1}{2}}z\}}{(k^2-\kappa^2)^{\frac{1}{2}}}J_0\{\kappa(x^2+y^2)^{\frac{1}{2}}\}\kappa\,d\kappa$$

$$=i2\pi\frac{e^{-ikr}}{r},\qquad z>0, \tag{2.60}$$

since the negative imaginary root (2.10) must always be taken for outgoing waves. The partial derivatives in (2.59) can then be evaluated simply from the fact that

$$\frac{\partial}{\partial z}\left(\frac{e^{-ikr}}{r}\right)=\frac{e^{-ikr}}{r}\left(-ik-\frac{1}{r}\right)\sin\theta_y\cos\phi_y, \tag{2.61a}$$

$$\frac{\partial}{\partial x}\left(\frac{e^{-ikr}}{r}\right)=\frac{e^{-ikr}}{r}\left(-ik-\frac{1}{r}\right)\sin\theta_y\sin\phi_y, \tag{2.61b}$$

in terms of the spherical coordinates (r, θ_y, ϕ_y) in which θ_y and ϕ_y denote the polar angles in Fig. 2.9 obtained when the y-axis, rather than the z-axis, is taken as the polar axis.

When the magnetic field is resolved into its spherical components $H_{\theta_y}{}^{\mathrm{d}}$, $H_{\phi_y}{}^{\mathrm{d}}$, $H_r{}^{\mathrm{d}}$ by the component transformation

$$H_{\theta_y}{}^{\mathrm{d}}=H_x{}^{\mathrm{d}}\cos\theta_y\sin\phi_y-H_y{}^{\mathrm{d}}\sin\theta_y+H_z{}^{\mathrm{d}}\cos\theta_y\cos\phi_y, \tag{2.62a}$$

$$H_{\phi_y}{}^{\mathrm{d}}=H_x{}^{\mathrm{d}}\cos\phi_y-H_z{}^{\mathrm{d}}\sin\phi_y, \tag{2.62b}$$

$$H_r{}^{\mathrm{d}}=H_x{}^{\mathrm{d}}\sin\theta_y\sin\phi_y+H_y{}^{\mathrm{d}}\cos\theta_y+H_z{}^{\mathrm{d}}\sin\theta_y\cos\phi_y, \tag{2.62c}$$

it is found from (2.59) to reduce to

$$H_{\theta_y}{}^{\mathrm{d}}(r,\theta_y,\phi_y)=0, \tag{2.63a}$$

$$H_{\phi_y}{}^{\mathrm{d}}(r,\theta_y,\phi_y)=\frac{Ib}{4\pi}\frac{e^{-ikr}}{r}\left(ik+\frac{1}{r}\right)\sin\theta_y, \tag{2.63b}$$

$$H_r{}^{\mathrm{d}}(r,\theta_y,\phi_y)=0, \tag{2.63c}$$

in complete agreement with the results obtained by the usual method based on the use of vector potentials. The corresponding electric field can then be

† A. Erdelyi (ed.), *Tables of integral transforms*, *vol* II, McGraw-Hill, 1954; p. 9, eqn (26).

obtained from the system equation (2.3*b*) by calculating the curl of this magnetic field.

It is evident from (2.63) that the angular variation in the far field is just $\sin\theta_y$, which is the well known radiation pattern of a Hertzian dipole.

2.8. Edge behaviour[7]

A presumption that is nearly always present in the mathematical formulation of source synthesis problems is that there are no physical limitations, in principle, on the admissible behaviour of the source itself. At the edges of the source the behaviour most often assumed is a pedestal; i.e. a function that approaches a finite non-zero value at the edges and that drops discontinuously to zero outside. The Hertzian dipole is an example of a hypothetical source that terminates in a pedestal at all four of its edges. But it will be shown here that, contrary to popular opinion, there is almost no arbitrariness at all in the behaviour that physically realizable sources can have at their edges. The pedestal, in particular, is not realizable physically, although it will be seen that it can be approximated as closely as desired for the component perpendicular to the edge. For the component parallel to the edge, however, the pedestal cannot even be approximated because the parallel component must always vanish at least linearly.

The role of edge behaviour in source synthesis was first recognized clearly by Taylor† in his monumental paper on the synthesis of narrow beam sources. With prophetic insight he hypothesized that the behaviour of any source at its edges be described by some power α of distance from the edges. He then showed that an essential analytic property of any pattern space factor is the asymptotic location of its remote zeros, and that this, in turn, is determined solely by his hypothesized behaviour of the source at its edges. He assumed that the exponent α could be any real number greater than -1, with all non-negative values regarded as being realizable physically. From this continuum of possible values he chose $\alpha=0$ (a pedestal) as being the best and the one to be sought in practice, on the grounds that it provides the greatest number of central zeros by which to control the shape of the pattern space factor in its visible region. Although this choice of α is now known to be unrealizable physically, an edge condition will be derived here that shows that his hypothesis on the *form* of the edge behaviour does, indeed, appear to be correct. Thus, all of the mathematical machinery that he developed on the relationship between the edge behaviour of the source and the zeros of the pattern space factor remains intact.

(i) *Exact edge conditions*

An antenna aperture has no independent physical identity apart from the antenna structure as a whole. Any physical description of the aperture

† T. T. Taylor. Design of line-source antennas for narrow beam-width and low side lobes. *IRE Trans. Antennas Propagat.* **AP-3,** pp. 16–28, Jan. 1955.

itself, therefore, would be incomplete without including a detailed description of the way in which the conducting planar sheet in which it is located is joined mechanically at the aperture edges to the rest of the antenna structure. The most general description is that of a conducting wedge formed by the intersection of the conducting planar sheet with the interior conducting walls of the exciting cavity. This is illustrated in Fig. 2.10. An essential part of the problem of physical realizability is to determine the realizable range of values of the edge angle ξ at which the interior and exterior walls of the antenna can intersect physically. This will be determined as follows.

The dominant behaviour of the field functions in the immediate neighbourhood ($r \to 0$) of the edge of a perfectly conducting wedge has been

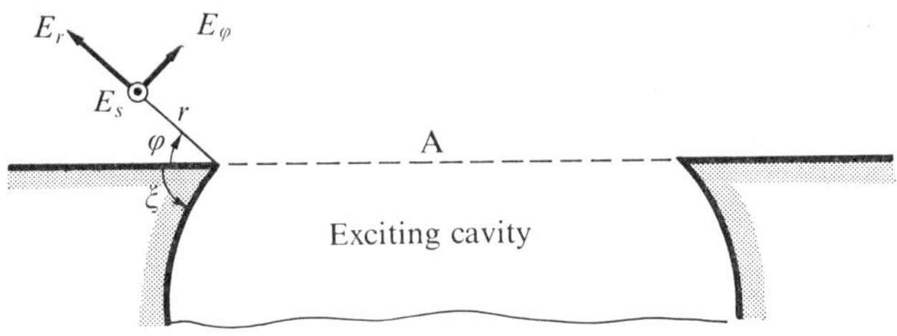

FIG. 2.10. Hypothetical cross-section of a planar antenna in the neighborhood of its aperture A. The edges of the aperture are assumed to be formed by the intersection of the exterior conducting planar surface with the interior conducting walls of the exciting cavity. The intersection forms a conducting wedge of angle ξ.

investigated by various authors in a number of different ways with essentially identical results:

$$E_s \sim C_1 r^q \sin q\phi, \tag{2.64a}$$

$$E_r \sim C_2 q r^{q-1} \sin q\phi, \tag{2.64b}$$

$$E_\phi \sim C_2 q r^{q-1} \cos q\phi, \tag{2.64c}$$

$$H_s \sim -i\omega\epsilon C_2 r^q \cos q\phi + C_3, \tag{2.64d}$$

$$H_r \sim (C_1/-i\omega\mu) q r^{q-1} \cos q\phi, \tag{2.64e}$$

$$H_\phi \sim (C_1/i\omega\mu) q r^{q-1} \sin q\phi, \tag{2.64f}$$

where the constants C_1, C_2, and C_3 are independent of the polar coordinates (r, ϕ) defined in Fig. 2.10 and also independent of the distance s along the edge. The parameter q is determined by the wedge angle ξ,

$$q = \frac{\pi}{2\pi - \xi}. \tag{2.65}$$

For a wedge with a straight edge the results (2.64) can be established from the fact that the field behaviour in a vanishingly small region of space is essentially the same as if the region were of fixed size and the wavelength approached infinity. The electric field then behaves essentially like a static field, which can be obtained as Maxwell did† from the static electric field above a conducting plane by conformal transformation of the plane into a wedge. The magnetic field that couples to it then follows directly from the system equations (2.3). Similar results have been obtained by Meixner‡ for a wedge with a smoothly curved edge by assuming a solution to the wave equation in the form of a power series in r and introducing an additional hypothesis that the fields must be square-integrable in the neighbourhood of the edge. The rigour of Meixner's power-series assumption has been questioned by Bouwkamp§ and by Heins and Silver,‖ however, the latter of whom reformulated the problem in quite a different way for the special case of $\xi = 0$ and again obtained a solution that is consistent with (2.64). Thus, although a conclusive proof of the validity of (2.64) is yet to be established for the general case of a wedge whose edge is curved, it does appear to be a reasonably safe conjecture.

From (2.64) it is evident that the dominant field behaviour near the edge of a conducting wedge is determined solely and completely by the value of q. And from (2.65) it is evident further that the value of q is determined by the wedge angle ξ. The value of q can be no less than $\frac{1}{2}$ since the smallest value possible for the wedge angle ξ is zero (a knife edge). Furthermore the value of q can be no greater than unity since a wedge angle in excess of 180° would cause the interior structure of the antenna to protrude through the aperture, thereby destroying the planar character of the aperture. Thus the value of q is limited by purely geometric considerations to the range

$$\tfrac{1}{2} \leqslant q \leqslant 1. \qquad (2.66)$$

Anywhere within that range the fields in the neighbourhood of the aperture edge would always be *mathematically realizable* in the sense that they satisfy the system equations and all boundary conditions. For any value less than unity, however, the radial and angular field components would all have to become infinite at the edge. Since infinite fields cannot exist physically it is concluded that there is but one *physically realizable* value of q, namely, $q = 1$. It then follows from (2.65) that the only physically realizable wedge angle is $\xi = 180°$. That is, *the edges of all physical apertures must be smooth*, not sharp in the exact geometric sense. This is in complete accord with experience.

† J. C. Maxwell. *A treatise on electricity and magnetism, vol.* 1, reprinted by Dover, New York; p. 295, 1954.

‡ J. Meixner. The behavior of electromagnetic fields at edges. Res. Rept. EM-72, New York University, Dec. 1954; reprinted in *IEEE Trans. Antennas Propagat.* **AP-20**, pp. 442–6, July, 1972.

§ C. J. Bouwkamp. Diffraction theory. *Rep. Prog. Phys.* **17**, pp. 35–100, 1954; p. 45.

‖ A. E. Heins and S. Silver. The edge conditions and field representation theorems in the theory of electromagnetic diffraction. *Proc. Camb. phil. Soc.* **51**, pp. 149–61, 1955.

The geometrical shape that the cross-section of a physically realizable planar antenna must have at the edges of its aperture is becoming clearer: the interior wall of the cavity exciting the aperture must be joined to the exterior wall in such a way that the transition from one wall to the other is continuous and has a continuous derivative (slope). That guarantees smoothness. The second derivative (curvature), however, must change discontinuously at the aperture edge from the zero value that it had on the plane to some non-zero value on the interior wall of the cavity. For if it did not the joint would be merely a continuation of the plane and not the edge of an

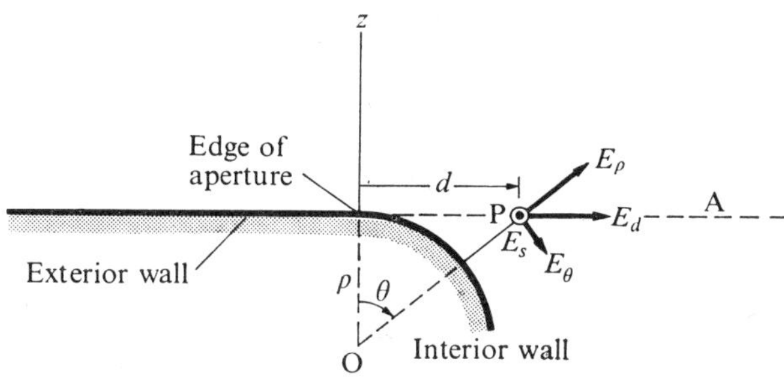

FIG. 2.11. Cross-section of any physically realizable planar antenna in the neighborhood of one edge of its aperture.

aperture at all. It is concluded, therefore, that the dominant property determining the shape of the interior wall in the immediate neighbourhood of an aperture is its radius of curvature ρ. This is illustrated in Fig. 2.11.

The behaviour of the electric field components tangential to the aperture (i.e. the aperture distribution) at a point P approaching the edge of the aperture can now be determined from the geometry of Fig. 2.11 as the angle θ vanishes. In a vanishingly small neighbourhood of the aperture edge the height of the point P above the interior wall becomes vanishingly small. The electric field behaviour then becomes essentially the same as that near a conducting plane; i.e. its tangential components vanish linearly with height above the conductor and its normal component approaches a constant value. Thus, the two components E_s and E_θ tangential to the curved interior wall must approach zero linearly with height as $C_1\{(\rho/\cos\theta)-\rho\}$, and the component E_ρ normal to the interior wall must approach a constant value C_2. As a consequence the components E_s and E_θ must each approach zero as $\frac{1}{2}C_1\rho\theta^2$, and the component E_d lying in the aperture plane in a direction perpendicular to the edge of the aperture must then approach just the projection $C_2\theta$ of E_ρ onto the plane of the aperture. Since θ approaches the value d/ρ the two components E_d and E_s that lie in the aperture plane must behave at the edges of the aperture as

$$E_d \sim (C_2/\rho)d, \tag{2.67a}$$

$$E_s \sim (C_1/2\rho)d^2. \tag{2.67b}$$

Thus, it is concluded that the tangential aperture field component E_d perpendicular to the edge of the aperture must vanish at least linearly with the distance d from the edge and that the component E_s parallel to the edge must vanish at least quadratically with the distance d. Stated another way:

In any planar aperture the tangential components of electric field (the aperture distribution) that are perpendicular or parallel to the edge of the aperture must vanish with distance d from the edge as d^α, where the dominant value of the edge exponent α is

$$\alpha_\perp = 1, \tag{2.68a}$$

$$\alpha_\parallel = 2, \tag{2.68b}$$

for the perpendicular and parallel components, respectively.

These constitute the exact edge conditions for planar sources. They are unique, with the possible exception of certain cases that are considered to be unlikely to occur in practice. These exceptional cases could arise only in the extraordinary event that the constants C_1 and C_2 in the lowest-order terms (2.64) should happen to be precisely zero. In such a case the edge behaviour would then be dominated by some higher power of r and the realizable values obtained for the exponent α would be greater (but never smaller) by some integer. The values given by (2.68) would then represent only a lower bound. But if C_1 and C_2 have any non-zero value, no matter how small, the above values of α will indeed be unique. Since it is highly improbable that the lowest-order terms would happen to be precisely zero for an actual antenna configuration, it is believed that this aperture edge condition is a fundamental property of almost all planar antennas.

(ii) *Approximation to the edge conditions*

It is now clear why the pedestal form of edge behaviour, on which most of the literature on aperture antenna theory has been based, can never be realized physically. Both of the electric field components lying in the plane of the aperture become tangential to the conducting interior wall upon approaching the edge of the aperture and, hence, must vanish right at the edge itself. Even though a pedestal cannot be realized exactly, however, it appears that the edge behaviour for the component E_d perpendicular to the edge can be made arbitrarily close to a pedestal by making the radius of curvature ρ of the conducting interior wall sufficiently small. This is evident from the observation that the slope C_2/ρ of the linear edge behaviour (2.67*a*) increases with decreasing radius ρ. Hence, the slope of the linear edge behaviour can be made to approximate the infinite slope of a pedestal as accurately as desired by making ρ sufficiently small. But the limiting case of infinite slope itself can never be attained in practice because it would require zero radius of curvature at the metallic joint. This would be the non-physical case of an infinitely sharp edge again.

In the case of the component E_s parallel to the aperture edge the edge behaviour cannot even be made to approximate a pedestal. This is evident from the observation that the slope $(C_1/\rho)d$ of its edge behaviour (2.67*b*) is always zero right at the edge $d = 0$, no matter how small (but non-zero) the radius of curvature ρ may be. On the other hand, the slope itself can always be made to approximate a pedestal (a finite discontinuity in slope) as accurately as desired by making ρ small enough, in precisely the same sense that the edge behaviour of E_d can be made to approximate a pedestal (a finite discontinuity in E_d). A pedestal for the slope of E_s would correspond to a linear edge behaviour of E_s itself. Thus, it appears that the edge behaviour of E_s can be made to approximate some linear function as accurately as desired.

It is concluded, therefore, that the dominant edge behaviour of E_d can be made arbitrarily close to a pedestal ($\alpha_\perp = 0$) and that the dominant edge behaviour of E_s can be made arbitrarily close to linear ($\alpha_\parallel = 1$) by making the radius of curvature ρ of the interior wall sufficiently small at the aperture edge. Regardless of how close the approximation might become, however, the actual physical values of the edge exponent α can never depart from those given by (2.68).

(iii) *Absolute minimum* α

Even in the mathematically realizable but physically unrealizable range $\frac{1}{2} \leqslant q < 1$, within which the field behaviour (2.64) would become infinite at the edges of the aperture, it should be noted that the two components E_d and E_s of the aperture distribution would still behave as some power of the distance r from the edge. The value of the exponent would be $q-1$ for the perpendicular component E_d and q for the parallel component E_s. These values, in turn, are determined uniquely by the wedge angle ξ at which the interior wall intersects the aperture plane. Since q has an absolute minimum value it follows that there must be an absolute minimum value for the edge exponents $\alpha_\perp$ and $\alpha_\parallel$ below which the edge behaviour would not even be mathematically realizable. The absolute minimum value of q is $\frac{1}{2}$, occurring when the wedge collapses to a knife-edge of angle $\xi = 0$. Hence, the absolute minimum values of α are

$$\alpha_{\perp\mathrm{min}} = -\tfrac{1}{2}, \tag{2.69a}$$

$$\alpha_{\parallel\mathrm{min}} = +\tfrac{1}{2}. \tag{2.69b}$$

These values of α have appeared frequently in the literature on electromagnetic theory. They are the values possessed by the edge behaviour of all of the classical solutions of planar diffraction problems (half-planes, circular apertures, etc.). They are also the values obtained by Meixner† from his edge condition for diffraction by apertures in planar screens. For quite different reasons they will be seen in section 5.5 to enter into the choice of constraining parameter for the case of strip sources and, hence, into the constrained synthesis solutions of Chapters 8 and 9.

(iv) *Effect on performance*

Edge behaviour, of itself, can have no direct effect whatsoever upon the overall performance (beam shape, sidelobe level, etc) of a planar source. This follows from the fact that edge behaviour is a phenomenon that occurs over only a vanishingly small portion of the aperture area. All of the performance properties are determined by the pattern space factor, and that, in turn, is determined by the integral (2.13) of the aperture distribution over the aperture area. But there can be no nonzero contribution to an integral over a region of zero area, providing that the integrand is integrable and contains no singularity functions (delta functions or their derivatives). The integrand of (2.13) will always be integrable at the edges for all values of α that exceed -1. Since the value of α can never be less than $-\frac{1}{2}$, from (2.69), it is clear that there can never be a contribution from the edges themselves.

On the other hand, edge behaviour might have an indirect effect upon performance to the extent that it influences the behaviour of the aperture distribution in the neighbourhood of the edges. It is only this indirect effect that can be of any practical significance. The extent of this effect upon Q-constrained optimum synthesis solutions is not yet known.

2.9. Physical antennas[7]

A more realistic model of slot or dipole antennas than the idealized one treated in section 2.5 must be devised now that it is known that all physically realizable aperture edges must be smoothly rounded instead of sharp. A sheet of zero thickness can never have rounded edges, hence it is concluded that the thickness of any physical conducting sheet must always be greater than zero, again in complete accord with experience. As a consequence any hole cut in a physical conducting sheet will create not just a single aperture but a *pair* of parallel apertures on either side of the sheet. This is illustrated for a slot antenna in Fig. 2.12a, in which the edge of the hole is assumed, for simplicity, to be rounded uniformly with a constant radius of curvature ρ. The open volume contained between the two parallel apertures constitutes a cavity that excites each of them from one side.

The dipole antenna that is complementary to the slot antenna in Fig. 2.12a must also have rounded edges (Fig. 2.12b) in order to be realizable physically.

† J. Meixner. Die kantenbedingung in der theorie der beugung elektromagnetischer wellen an vollkommen leitenden ebenen schirmen. *Ann. Phys.*, **6**, pp. 2–9, 1949.

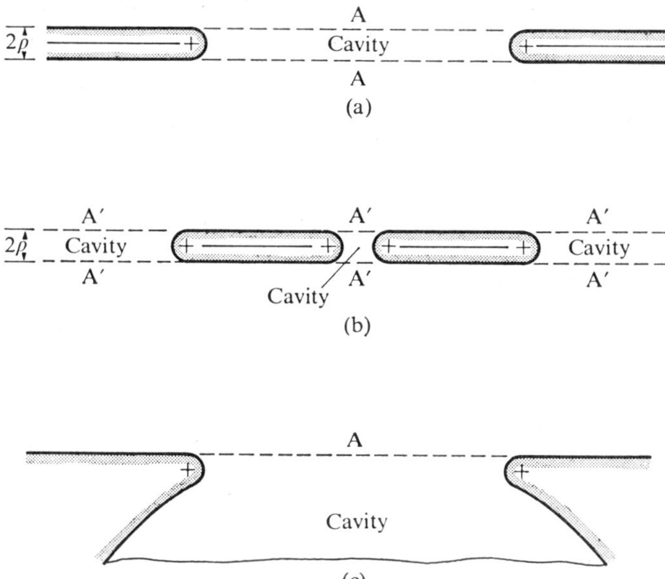

FIG. 2.12. Cross-section of the three physical types of planar antenna possible: (a) slot, (b) dipole, (c) arbitrary cavity-excitation of a single aperture.

Hence it, too, must have a pair of parallel apertures on either side of an open cavity. Here the pair of apertures A' is roughly the complement of the conducting sheet containing the pair of apertures A of the slot antenna (the small difference arises from the narrow gap needed in the centre of the dipole in order to drive it), which means that the cavity between them must extend to infinity.

Slot and dipole antennas are of limited practical utility because they provide relatively little control over the design. Their principal means of control lies in the choice of the size and shape of the aperture pair. A much more useful configuration is the third one illustrated in Fig. 2.12c in which just a single aperture is excited by a conducting cavity of arbitrary size and shape. The cavity can be filled with a material medium whose constitutive parameters are arbitrary functions of position. It is this third type of planar antenna that was envisioned in section 1.3 for the formulation of the general antenna synthesis problem. The arbitrariness of the constitutive parameters provides almost unlimited flexibility of design.

It is important to recognize that all of the three possible types of planar antenna illustrated in Fig. 2.12 have a common physical mechanism of radiation. In all three cases each aperture is excited from just one side by a cavity. Hence there can be an actual flow of power through the aperture. This is distinctly different from the radiation mechanism hypothesized in section 2.5 for idealized slot or dipole antennas of zero thickness, for which no power at all could flow through the aperture itself. The presence of a cavity also serves the important function of providing a realistic means for connecting a transmission line to the antenna, whereas none at all is available for idealized slot or dipole antennas (a coaxial cable of zero thickness is hardly realistic). Thus, by choosing the model for a slot or dipole antenna to have nonzero thickness, no matter how small that thickness may be, it will be guaranteed to possess the properties required for physical realizability.

Now that a more realistic model has been obtained it is of interest to reconsider the question of edge behaviour in the limiting case of a physical slot or dipole antenna whose thickness 2ρ is reduced exactly to zero. As long as ρ remains greater than zero the two tangential components of electric field in each one of the pair of parallel apertures will vanish at the edges in accordance with (2.67). One might suppose, then, that the same edge behaviour would continue to persist in the limiting case of exactly zero thickness. It need not, however, because the edge behaviour derived here was based upon the fact that fields in a sufficiently small region of space near a conducting curved surface behave essentially like those near a conducting plane. Zero thickness implies zero radius of curvature at the edge, which means that a field near such a curved surface would fail to behave as it would near a plane no matter how near one approached the surface. So one cannot say, even in principal, what the edge behaviour would become if the thickness of the sheet were to be reduced exactly to zero.

CHAPTER 3

THE FUNDAMENTAL PRINCIPLE OF PLANAR SOURCES

THERE is a widely-held belief that the radiation pattern of an antenna determines only its radiative properties and that to describe its reactive properties one has no choice but to work directly with the fields at the radiating surface. Radiation resistance, for example, is commonly calculated from the radiation pattern, whereas input reactance is almost invariably based upon the fields at the surface of the antenna. It is largely because of this presumed failure to account for energy storage that the radiation pattern has always assumed a secondary role in antenna theory. In general this failure may be real, as no proof to the contrary is known to the author. But in the case of planar antennas the source and its pattern space factor are each derivable from the other through the Fourier transform (2.12). Hence the question arises as to whether or not it might be possible to base the entire theory of planar sources upon the radiation pattern instead of upon the source itself. Although this might at first appear to be overly optimistic it does, indeed, prove to be the case. The radiation pattern will be seen to possess one of the most severe constraints that mathematics can impose, in that the pattern space factor is an analytic function of the three complex variables k_x, k_y, and k. For any realizable radiation pattern the pattern space factor can always be continued analytically from its visible region of radiated power into the invisible region of reactive power. This analytic property of the pattern space factor, together with the Fourier transform relationship between the pattern and aperture functions, provides the key to the theory of realizable planar source synthesis. By expanding the visible part of the pattern space factor in some complete set of aperture-limited functions the analytic continuation from the visible to the invisible region will be obtained automatically.

The central and unifying role of the pattern space factor will be stated explicitly as a general principle. A number of consequences that are crucial to planar source synthesis will then be derived. One is an analytic extension of the classical Poynting vector method, by means of which input reactance can be determined as well as radiation resistance from a knowledge of the radiation pattern alone. Another is a logically consistent formulation for the stored magnetic and electric energies that determine the quality factor Q of planar sources. The new principle is the cornerstone of the theoretical foundations on which the synthesis of planar antenna sources will be based.

3.1. Statement of the principle[2]

The visible part of the pattern space factor has been identified in (2.39) with the conventional radiation pattern but the invisible part is still left undetermined. It is frequently claimed that any functional form can be assigned to the invisible part without affecting the visible part. In actual fact, however, the invisible part can never be specified independently of the visible part because of a rigid interdependence that is imposed by analyticity. Failure to recognize this essential point is responsible for much of the misunderstanding that continues to persist regarding the reactive properties of radiating sources.

Analyticity of the pattern space factor can be established directly from its integral representation in terms of the real impulse-response function $\hat{\mathscr{E}}(x, y, 0, \tau)$ of the real variables x, y, and τ. From the definition (2.2*a*) the complex field function $\mathbf{E}(x, y, z, k)$ at any point in space can be written

$$\mathbf{E}(x, y, z, k) = \int_{r/c}^{\infty} \hat{\mathscr{E}}(x, y, z, \tau)e^{-ikc\tau}\,d\tau, \tag{3.1}$$

since the impulse response at any distance r from the input of any causal system must be zero prior to the time r/c after the occurrence of the input impulse. Hence the rectangular components (2.13) of the pattern space factor become

$$F_x(k_x, k_y, k) = \iint_A \int_{r/c}^{\infty} \hat{\mathscr{E}}_x(x, y, 0, \tau)e^{i(k_x x + k_y y - kc\tau)}\,dx\,dy\,d\tau, \tag{3.2a}$$

$$F_y(k_x, k_y, k) = \iint_A \int_{r/c}^{\infty} \hat{\mathscr{E}}_y(x, y, 0, \tau)e^{i(k_x x + k_y y - kc\tau)}\,dx\,dy\,d\tau. \tag{3.2b}$$

This integral representation for the pattern space factor is a complex function of the three complex variables k_x, k_y, and k. The analytic properties of functions of several complex variables are far more difficult to establish and to use than those of a single complex variable, in general. But for a certain class of functions of several complex variables its analytic properties reduce simply to those of each complex variable separately. This is a consequence of a theorem attributed to F. Hartogs,† which states that a function of several complex variables will be analytic in any domain for which it is an analytic function of each complex variable separately when all of the others are held fixed at any given point within that domain. The triple integral (3.2) is a function of that particular class; it satisfies the Cauchy-Riemann conditions for each of the three complex variables k_x, k_y, and k separately in any domain

† S. Bochner and W. T. Martin. *Several complex variables*, Princeton University Press, Princeton, New Jersey, 1948; p. 140.

for which the integral exists. The domain of analyticity will then be the domain of summability of the triple integral. In either of the complex k_x or k_y planes the integral is summable everywhere except at the point at infinity, because of the fact that the aperture A has finite dimensions and because all physical impulse-response functions must have finite values everywhere in space and in time. And in the complex k plane it is summable everywhere below some positive imaginary part whose value is determined by the losses in the system. Thus, each of the pattern space factor components (3.2) is an entire analytic function (of exponential type†) of k_x and k_y and an analytic function of k in the lower half-plane.‡

Analyticity in the lower half of the complex k, or ω, plane is a requirement that arises from the temporal behaviour of impulse-response functions. Hence it is common to all physically realizable systems, not just to radiating systems. But analyticity in the complex k_x and k_y planes, which has spatial rather than temporal origins, may be unique to radiating systems.

An essential consequence of analyticity is that any piece of an analytic function, no matter how small, determines all the rest completely and uniquely by analytic continuation. By a piece of an analytic function of a single complex variable is meant the values of the function over any arbitrarily small arc, or even over any infinite number of distinct points, that are contained within the domain of analyticity.§ In the case of the pattern space factor (3.2) the arc could be any nonzero segment of the real axis in each of the complex k_x, k_y, and k planes. At any given frequency, therefore, it could be any nonzero area in the (k_x, k_y) plane. In particular, it could be the visible region $k_x^2+k_y^2 < k^2$ defining the radiation pattern, or any part thereof.

Since any piece of the pattern space factor determines the rest completely by analytic continuation, including its values everywhere along the real axes, and since its values along the real axes determine the field functions (2.16) everywhere in space, one is led to the following remarkable principle:

All of the properties of any planar source can be determined completely from any piece of its pattern space factor.‖

Because of the key role that it will be seen to play in the theory of planar source synthesis it will be referred to henceforth as the *fundamental principle of planar sources.*

Not all analytic functions represent realizable pattern space factors. The only ones that do are those that can be expressed as the Fourier transform (3.2) of some causal impulse-response function that vanishes outside a given area A and that behaves properly at the edges. The construction of

† R. E. A. C. Paley and N. Wiener. *Fourier transforms in the complex domain*, Amer. Math. Soc. Colloq. Pubs, vol. 19, Providence, Rhode Island, 1934; p. 13.

‡ E. T. Whittaker and G. N. Watson. *Modern analysis*, Cambridge University Press, 1952; pp. 92–3.

§ K. Knopp. *Theory of functions, Part I.* Dover Publications, New York, 1945; p. 87.

‖ The principle was stated originally in terms of TE and TM partial pattern functions.

such functions for rectangular and for circular sources will be treated in detail in Chapter 4.

The analytic character of the pattern space factor and the consequences thereof have not always been appreciated. An extreme example, but one that is important because it is so often misunderstood, is the hypothetical case in which the pattern space factor is assumed to be precisely zero in its invisible region. For such a case the pattern space factor, and hence the source that produces it, would have to be zero everywhere.

As a word of caution, one should not expect to be able to determine experimentally the reactive properties of planar sources by means of analytic continuation of their measured far-field patterns. The measurement accuracy required to continue an experimentally determined piece of an analytic function without excessive error propagation is generally prohibitive.

3.2. Analytic extension of the Poynting vector method[2]

The earliest method used to compute radiation resistance of an antenna was to compute the power radiated from the antenna by integrating the Poynting vector of the far-field over an infinitely large sphere and then to divide it by the mean-squared current at the input terminals.† This is the well-known Poynting vector method. In its original form it was restricted to a determination of just the resistive part of the input impedance because of the fact that the power in the far-field is purely radiative. It will now be shown that for any planar source the reactive power can be determined from the far-field by means of analytic continuation of the pattern space factor from its visible into its invisible region. As a consequence, the complete input impedance can be determined, exactly and uniquely, by extending the Poynting integration from the interior of the circle $k_x^2+k_y^2 = k^2$ to the entire wavenumber plane. This constitutes an *analytic extension of the Poynting vector method.*

The complex power at any planar source is represented by the surface integral of the complex Poynting vector over the aperture A,

$$-\tfrac{1}{2}\int\limits_A \mathbf{E}\times\mathbf{H}^*.\mathbf{n}\,da = \tfrac{1}{2}\int\limits_{-\infty}^{\infty}\int\limits_{-\infty}^{\infty}(E_x H_y^* - E_y H_x^*)\,dx\,dy, \tag{3.3}$$

where the asterisk denotes the complex conjugate. The unit normal $\mathbf{n}$ is directed outward (negative z direction) from the volume containing the fields, and the boundary condition on E_x and E_y requires that the integration be extended from just the aperture to the entire (x, y) plane. Expressing each of the six field components in terms of the pattern space factor by (2.16) and then interchanging the order of integration one finds that each of the integrals

† B. van der Pol. On the wavelengths and radiation of loaded antennae, *Proc. Phys. Soc.* (*London*), **29**, pp. 269–89, June 1917.

over x and y reduces to 2π times a delta function. Thus, the sextuple integral collapses to just a double integral over the wavenumber plane,

$$-\tfrac{1}{2}\int_{A}\mathbf{E}\times\mathbf{H}^{*}.\mathbf{n}\,da = \frac{1}{(2\pi)^{2}2kZ_{0}}\times$$

$$\times\int_{-\infty}^{\infty}\int_{-\infty}^{\infty}\frac{(k^{2}-k_{y}^{2})\,|F_{x}|^{2}+(k^{2}-k_{x}^{2})\,|F_{y}|^{2}+2k_{x}k_{y}\operatorname{Re}F_{x}F_{y}^{*}}{\{(k^{2}-k_{x}^{2}-k_{y}^{2})^{\frac{1}{2}}\}^{*}}\,dk_{x}\,dk_{y}. \quad (3.4)$$

This is a complete formulation for the Poynting integration of the pattern space factor. Its real part, arising from the visible region $k_{x}^{2}+k_{y}^{2} < k^{2}$ alone, represents the radiative power through the aperture,

$$\operatorname{Re}\left(-\tfrac{1}{2}\int_{A}\mathbf{E}\times\mathbf{H}^{*}.\mathbf{n}\,da\right) = \frac{1}{(2\pi)^{2}2kZ_{0}}\times$$

$$\times\iint_{k_{x}^{2}+k_{y}^{2}<k^{2}}\frac{(k^{2}-k_{y}^{2})\,|F_{x}|^{2}+(k^{2}-k_{x}^{2})\,|F_{y}|^{2}+2k_{x}k_{y}\operatorname{Re}F_{x}F_{y}^{*}}{(k^{2}-k_{x}^{2}-k_{y}^{2})^{\frac{1}{2}}}\,dk_{x}\,dk_{y}. \quad (3.5)$$

This is exactly equivalent to integration of the radiation pattern over a hemisphere on one side of the aperture plane.

All the rest of integral (3.4), obtained from analytic continuation of the pattern space factor into its invisible region, represents the reactive power at the aperture:

$$\operatorname{Im}\left(-\tfrac{1}{2}\int_{A}\mathbf{E}\times\mathbf{H}^{*}.\mathbf{n}\,da\right) = \frac{1}{(2\pi)^{2}2kZ_{0}}\times$$

$$\times\iint_{k_{x}^{2}+k_{y}^{2}>k^{2}}\frac{(k_{y}^{2}-k^{2})\,|F_{x}|^{2}+(k_{x}^{2}-k^{2})\,|F_{y}|^{2}-2k_{x}k_{y}\operatorname{Re}F_{x}F_{y}^{*}}{(k_{x}^{2}+k_{y}^{2}-k^{2})^{\frac{1}{2}}}\,dk_{x}\,dk_{y}, \quad (3.6)$$

since the square root in the denominator of (3.4) must take the purely imaginary value (2.10) in the invisible region.

These expressions for the radiative and reactive powers of planar sources will now be used to compute the input impedance of narrow rectangular dipole antennas whose pattern is assumed to be that produced by a sinusoidal distribution of current. The result will be seen to agree precisely with the usual value for impedance as obtained by the conventional induced-e.m.f. method.

The distribution of current along the length of a narrow dipole antenna operating near half-wave resonance is known to become sinusoidal as the width vanishes. Assuming that it is uniform across the width, the electric surface current density on one side of the dipole is represented by

$$K_y(x, y, k) = \frac{I_{\max}}{2a} \sin k(\tfrac{1}{2}b - |y|), \tag{3.7}$$

where $I_{\max}$ represents the maximum current on a dipole of width a along the x-axis and length b along the y-axis. The current density on each side of the dipole is determined by only half of $I_{\max}$ because the total current divides equally between the two sides.

The magnetic field supported by this surface current density has but a single tangential component $H_x^{\mathrm{d}}(= \mathbf{n} \times \mathbf{K})$. Hence the pattern space factor (2.49) must also have but a single component, which is found to be

$$F_x^{\mathrm{d}}(k_x, k_y, k) = kZ_0 I_{\max} \frac{\sin \frac{1}{2}k_x a}{\frac{1}{2}k_x a} \times \frac{\cos \frac{1}{2}k_y b - \cos \frac{1}{2}kb}{k^2 - k_y^2}. \tag{3.8}$$

The complex power at the dipole surface can now be represented by the complex conjugate† of the weighted volume (3.4) under the squared magnitude of the pattern space factor in the (k_x, k_y) plane,

$$-\tfrac{1}{2} \int_A \mathbf{E} \times \mathbf{H}^* . \mathbf{n} \, da =$$

$$= \frac{kZ_0 |I_{\max}|^2}{8\pi^2} \int_{-\infty}^{\infty} \int_{-\infty}^{\infty} \left(\frac{\sin \frac{1}{2}k_x a}{\frac{1}{2}k_x a} \right)^2 \frac{(\cos \frac{1}{2}k_y b - \cos \frac{1}{2}kb)^2}{(k^2 - k_y^2)(k^2 - k_x^2 - k_y^2)^{\frac{1}{2}}} \, dk_x \, dk_y. \tag{3.9}$$

Its real part, representing the time-average power radiated by the dipole antenna, is given by the volume within the visible region. And its imaginary part, representing the net time-average reactive power produced by the difference between the time-average magnetic and electric energies stored everywhere about the dipole antenna, is given by the volume within the invisible region. Within the strip $k_y^2 < k^2$ the net reactive power is positive (excess stored magnetic energy), whereas outside it is negative (excess stored electric energy). This is illustrated by the power density plot shown in Fig. 3.1 for the complementary case of a narrow half-wave slot antenna of vanishing width with applied voltage V_0 at the centre. The power density plot for the planar dipole would be identical to that of the complementary slot except for a constant scale factor and a sign reversal over the invisible region.

Power density plots like those of Fig. 3.1 provide an interesting geometrical interpretation of the phenomenon of self-resonance of idealized planar slot or dipole antennas. The net reactive power is represented by the volume under

† Magnetic and electric stored energies interchange for complementary sources, from (2.48).

the part of the power density surface that lies in the invisible region $k_x^2+k_y^2 > k^2$. Within the strip $k_y^2 < k^2$ that volume will always be of one sign, negative for slot antennas and positive for dipole antennas, whereas outside of the strip it will always be of the opposite sign. When these two volumes are equal the net reactive power will be zero, indicating that the input reactance of the antenna is zero. Hence, self-resonance occurs when the total (algebraic) volume under the invisible part of the power density surface is exactly zero.

The radiative and reactive powers on one side of the dipole antenna are represented by the real and imaginary parts, respectively, of (3.9). Thus, the radiative power is obtained from the visible region only and the reactive power from the invisible region only. As the width a approaches zero they approach the values

$$\mathrm{Re}\left(-\tfrac{1}{2}\int_A \mathbf{E}\times\mathbf{H}^*.\mathbf{n}\,da\right) =$$

$$= \frac{kZ_0\,|I_{\max}|^2}{2\pi^2}\int_0^k \frac{(\cos\frac{1}{2}k_y b-\cos\frac{1}{2}kb)^2}{k^2-k_y^2}\times$$

$$\times\left\{\int_0^{(k^2-k_y^2)^{\frac{1}{2}}}\left(\frac{\sin\frac{1}{2}k_x a}{\frac{1}{2}k_x a}\right)^2\frac{dk_x}{\{(k^2-k_y^2)-k_x^2\}^{\frac{1}{2}}}\right\}dk_y \doteqdot$$

$$\doteqdot \frac{Z_0\,|I_{\max}|^2}{2\pi^2}\,\frac{\pi}{2}\int_0^1 \frac{(\cos\frac{1}{2}kb\eta - \cos\frac{1}{2}kb)^2}{1-\eta^2}\,d\eta, \qquad (3.10)$$

$$\mathrm{Im}\left(-\tfrac{1}{2}\int_A \mathbf{E}\times\mathbf{H}^*.\mathbf{n}\,da\right) =$$

$$= \frac{kZ_0\,|I_{\max}|^2}{2\pi^2}\left[\int_0^k \frac{(\cos\frac{1}{2}k_y b - \cos\frac{1}{2}kb)^2}{k^2-k_y^2}\times\right.$$

$$\times\left\{\int_{(k^2-k_y^2)^{\frac{1}{2}}}^{\infty}\left(\frac{\sin\frac{1}{2}k_x a}{\frac{1}{2}k_x a}\right)^2\frac{dk_x}{\{k_x^2-(k^2-k_y^2)\}^{\frac{1}{2}}}\right\}dk_y +$$

$$\left.+\int_k^{\infty}\frac{(\cos\frac{1}{2}k_y b - \cos\frac{1}{2}kb)^2}{k^2-k_y^2}\left\{\int_0^{\infty}\left(\frac{\sin\frac{1}{2}k_x a}{\frac{1}{2}k_x a}\right)^2\frac{dk_x}{\{k_x^2+(k_y^2-k^2)\}^{\frac{1}{2}}}\right\}dk_y\right] \doteqdot$$

$$\doteqdot \frac{Z_0\,|I_{\max}|^2}{2\pi^2}\int_0^{\infty}\frac{(\cos\frac{1}{2}kb\eta - \cos\frac{1}{2}kb)^2}{1-\eta^2}\left(\tfrac{3}{2}-C-\ln\tfrac{1}{2}ka\,|1-\eta^2|^{\frac{1}{2}}\right)d\eta, \qquad (3.11)$$

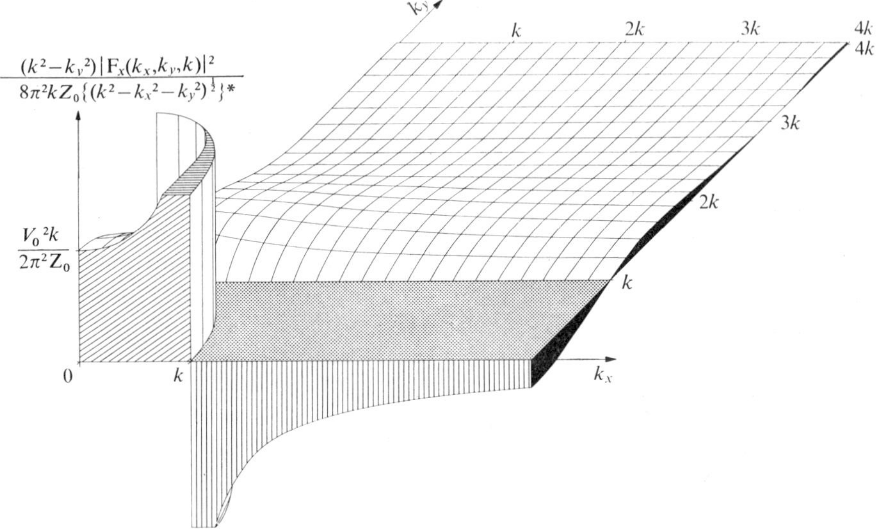

FIG. 3.1. One quadrant of the power density plot of a filamentary half-wave slot antenna with applied voltage V_0 at the center. Total radiated power on one side of the aperture plane is represented by the volume within the visible region $k_x{}^2+k_y{}^2 < k^2$, total reactive power by the algebraic volume outside.

where C denotes Euler's constant 0·57721···. The first of the two integrals on k_x in (3.11) arises from within the strip $k_y{}^2 < k^2$ and the second from the rest of the invisible region, but the two approach a common form, (A2.3) and (A2.4), for small values of a. Thus, as a approaches zero the two integrals can be combined into the single integral shown.

The radiation resistance and input reactance of the dipole antenna can now be calculated from twice the ratio of the complex power to the mean-squared value of the current $I_{\max} \sin \frac{1}{2}kb$ at the input terminals, the factor of two being necessary to account for the fact that power is supplied to both sides of the dipole:[4]

$$R = \frac{2\,\mathrm{Re}\left(-\frac{1}{2}\int\limits_A \mathbf{E}\times\mathbf{H}^*.\mathbf{n}\,da\right)}{\frac{1}{2}\,|\mathrm{I}_{\max}\sin\frac{1}{2}kb|^2} \doteqdot$$

$$\doteqdot \frac{Z_0}{2\pi\sin^2\frac{1}{2}kb}\{\mathrm{Cin}\,kb+(\mathrm{Cin}\,kb-\tfrac{1}{2}\,\mathrm{Cin}\,2kb)\cos kb-$$

$$-(\mathrm{Si}\,kb-\tfrac{1}{2}\,\mathrm{Si}\,2kb)\sin kb\}, \tag{3.12}$$

$$X = \frac{2\,\mathrm{Im}\left(-\frac{1}{2}\int\limits_A \mathbf{E}\times\mathbf{H}^*.\mathbf{n}\,da\right)}{\frac{1}{2}\,|\mathrm{I}_{\max}\sin\frac{1}{2}kb|^2} \doteqdot$$

$$\doteqdot \frac{Z_0}{2\pi\sin^2\frac{1}{2}kb}\Big\{\mathrm{Si}\;kb+(\mathrm{Si}\;kb-\tfrac{1}{2}\,\mathrm{Si}\;2kb)\cos kb+$$

$$+\left(\mathrm{Cin}\;kb-\tfrac{1}{2}\,\mathrm{Cin}\;2kb-\ln\frac{e^{\frac{3}{2}}b}{2a}\right)\sin kb\Big\}, \tag{3.13}$$

where the closed-form expression (A2.5) for the integral in (3.10), and (A2.6) and (A2.7) for the integral in (3.11), have been used. Note that the expression for radiation resistance becomes independent of the width a, while that for input reactance increases logarithmically with decreasing width. Numerical values for resistance and reactance in the neighborhood of resonance are shown in Fig. 3.2 as a function of length and width of the dipole at wavelength λ.

It is of interest to compare these expressions for the resistance and reactance of narrow planar dipole antennas obtained by analytic extension of the Poynting vector method with the conventional expressions for thin cylindrical dipole antennas obtained by the induced-e.m.f. method. The expressions for radiation resistance become identical to each other,† because of the fact that radiation resistance becomes insensitive to the cross-sectional dimensions of the antenna. And the expressions for input reactance also become identical‡ except for their dependence upon cross-section. The dependence upon cross-section cannot be compared directly because the cross-sectional geometries are different, the planar dipole being a planar strip of zero thickness and width a while the cylindrical dipole is a circular cylinder of radius r. But there still is a remarkable connection between the two. The expression for reactance of the planar dipole antenna of decreasing width a is seen to become identical to that for the cylindrical dipole antenna of radius

$$r = \frac{a}{e^{\frac{3}{2}}} \doteqdot \frac{a}{4{\cdot}48}. \tag{3.14}$$

Thus it is concluded that *the equivalent radius of a cylindrical dipole antenna is* $a/e^{\frac{3}{2}}$, in the sense that the value of the input impedance of cylindrical and planar dipole antennas with sinusoidal current distributions become identical as their cross-sectional dimensions approach zero.

This equivalent radius obtained from impedance considerations is different from, but close to, the equivalent radius $a/4$ obtained from static

† E. C. Jordan and K. G. Balmain. *Electromagnetic waves and radiating systems*, 2nd edn, Prentice-Hall, New Jersey, 1968; eqn. (14–21).

‡ Ibid., eqn. (14–24).

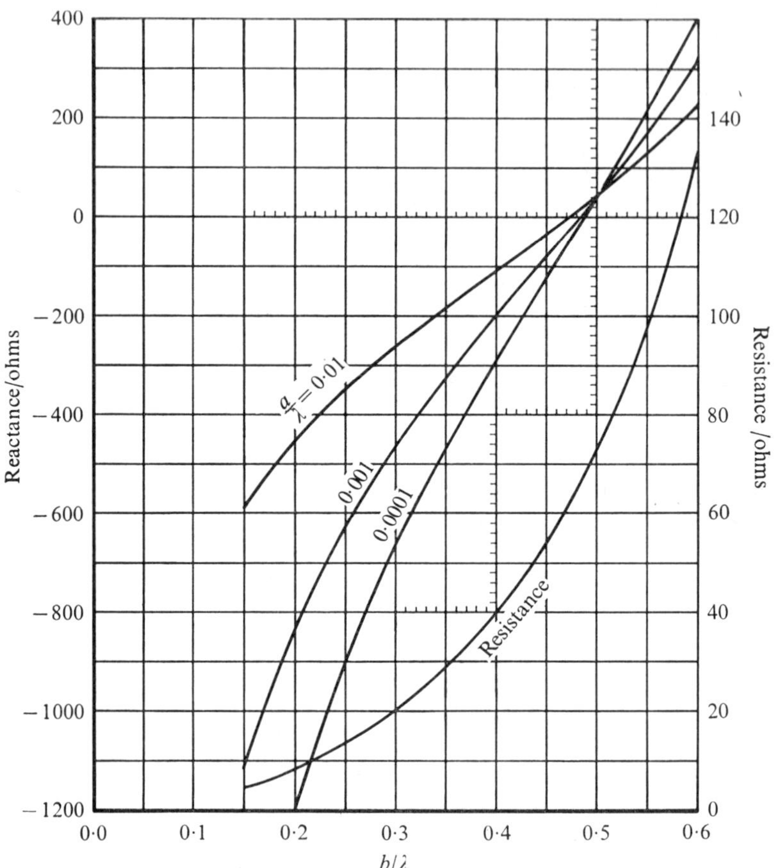

FIG. 3.2. Radiation resistance and input reactance of a planar dipole antenna of width a and length b as computed from the analytic extension of the Poynting vector method.

field considerations.† It is usually assumed that the static-field equivalence will lead to the correct impedance.‡ But it now appears that the static-field equivalence may not apply exactly to impedance calculations, although it does come within roughly 10 per cent of the value obtained here.

The analytic extension of the Poynting vector method as developed here has been limited exclusively to the case of planar sources because of the reversibility that is known to be possessed by the Fourier transform relationship between the pattern space factor and its source. It has been pointed out recently by Hill,§ however, that a reversible relationship exists also for the

† E. Hallen. *Electromagnetic theory*, John Wiley, New York, 1962, p. 62.

‡ Ibid, p. 462.

§ P. C. J. Hill. Analytic determination of dipole reactance by radiation-pattern integration. *Proc. Instn elect. Engrs*, **114,** pp. 853–8, July 1967.

case of circularly cylindrical sources (e.g. the cylindrical dipole antenna) and, hence, that the reactive power of circularly cylindrical sources can be determined by analytic continuation of the radiation pattern into the invisible region in much the same manner as was done here for planar sources. Hill's relationship between the pattern and aperture functions differs from that of planar sources in that his aperture function is a one-dimensional Fourier transform of the product of the pattern function by a Hankel function. In the particular case of a sinusoidal distribution along the length of a thin cylindrical dipole antenna his expression for reactive power obtained by integration over the invisible region of the pattern space factor leads to an expression for input reactance that is precisely the same as the conventional result obtained by the induced-e.m.f. method.

3.3. Mutual admittances in phased arrays

In principle the analytic extension of the Poynting vector method is equivalent to the induced-e.m.f. method. Instead of expressing the complex power at the source in terms of the source itself, as is done in the case of the induced e.m.f. method, it is expressed in an exactly equivalent form in terms of the pattern space factor. Thus the analytic extension of the Poynting vector method is not limited to just the determination of input impedances or admittances but can be used as well to determine mutual impedances or admittances between elements in phased arrays. The representation of mutual admittances in terms of pattern space factors has been found to be peculiarly suited to phased array analysis because it leads to a simple geometrical description of the way in which the pattern space factors of the individual elements of the array interact with the array factor to produce the active input admittance of each element when in the presence of all of the others. This will be demonstrated here. The effect of the array factor upon the input admittance of any element in the array turns out to be simply a weighting of the power density plot of a single element. This provides an important physical meaning that can be attributed to the visible and invisible regions of the array factor.

The essential features of the effects of mutual coupling between elements will be illustrated by treating the case of a uniformly spaced rectangular array of rectangular elements (slots) whose source has but a single component $E_x(x, y, 0, k)$. It will be assumed that the array lattice has $2M+1$ elements in the x direction and $2N+1$ elements in the y direction, and that each element is of width a in the x direction and length b in the y direction. The geometric centers of the individual elements are assumed to be located at $x = md_x$ and $y = nd_y$, as in Fig. 3.3, where d_x and d_y represent the distance between elements in the x and y directions, respectively, and where $m = 0, \pm 1, \pm 2, \cdots, \pm M$ and $n = 0, \pm 1, \pm 2, \cdots, \pm N$. The aperture distribution over the entire aperture plane $z = 0$ can then be expressed exactly as a linear

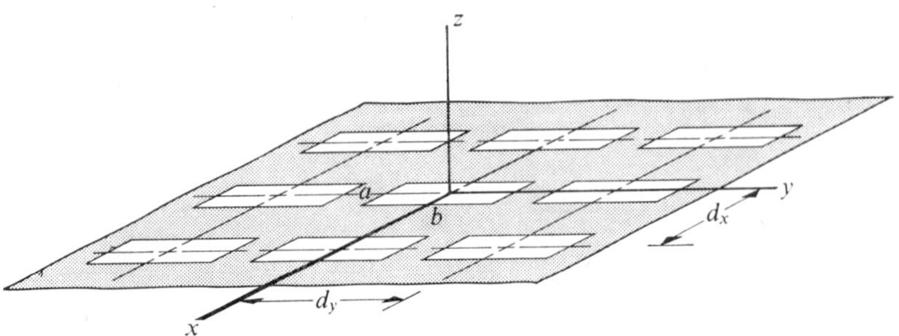

FIG. 3.3. Geometry of a phased array of rectangular slots. The slots are of width a and length b and are arrayed in a rectangular lattice with spacing d_x in the x direction and d_y in the y direction.

superposition of the individual distributions,

$$E_x(x, y, 0, k) = \sum_{m=-M}^{M} \sum_{n=-N}^{N} V_{mn} A_{xmn}(x-md_x, y-nd_y, k), \tag{3.15}$$

where $A_{xmn}(x-md_x, y-nd_y, k)$ denotes the aperture distribution produced in the mnth element alone per volt of excitation and V_{mn} denotes the exciting voltage of that element. Hence the pattern space factor (2.13) for the array as a whole can be written

$$F_x(k_x, k_y, k) = \sum_{m=-M}^{M} \sum_{n=-N}^{N} V_{mn} P_{xmn}(k_x, k_y, k) e^{i(mk_x d_x + nk_y d_y)}, \tag{3.16}$$

where the function

$$P_{xmn}(k_x, k_y, k) = \int_{-\frac{a}{2}}^{\frac{a}{2}} \int_{-\frac{b}{2}}^{\frac{b}{2}} A_{xmn}(\xi, \eta, k) e^{i(k_x\xi + k_y\eta)}\, d\xi\, d\eta \tag{3.17}$$

represents the pattern space factor of the mnth element alone.

The mutual admittances between array elements can now be obtained in terms of these pattern space factors (3.17) by calculating the complex power for the array as a whole. From (3.4) the complex power on one side of the aperture plane is seen to be represented by

$$-\tfrac{1}{2} \int_A \mathbf{E}\times\mathbf{H}^* . \mathbf{n}\, da = \sum_{m=-M}^{M} \sum_{n=-N}^{N} \sum_{p=-M}^{M} \sum_{q=-N}^{N} \tfrac{1}{2} V_{mn} V_{pq}^* Y_{mn,pq}^*, \tag{3.18}$$

where

$$Y_{mn,pq} = \frac{2}{(2\pi)^2 2kZ_0} \int_{-\infty}^{\infty} \int_{-\infty}^{\infty} \frac{(k^2-k_y^{\,2}) P_{xmn}^*(k_x, k_y, k) P_{xpq}(k_x, k_y, k)}{(k^2-k_x^{\,2}-k_y^{\,2})^{\frac{1}{2}}} \times$$
$$\times\, e^{i\{(p-m)k_x d_x + (q-n)k_y d_y\}}\, dk_x\, dk_y \tag{3.19}$$

defines the *mutual admittance* between the mnth and the pqth elements.

To obtain the input admittance Y_{mn} of the mnth element in the presence of all of the other excited elements note that an alternative expression for complex power (3.18) to the array as a whole can be written as

$$-\tfrac{1}{2}\int_A \mathbf{E}\times\mathbf{H}^*.\mathbf{n}\,da = \sum_{m=-M}^{M}\sum_{n=-N}^{N}\tfrac{1}{2}V_{mn}I_{mn}^* \tag{3.20}$$

in terms of the input current I_{mn} to the mnth element and the input voltage V_{mn} at that element. Since the current I_{mn} and voltage V_{mn} are related to the input admittance Y_{mn} of that element by

$$I_{mn} = V_{mn}Y_{mn} \tag{3.21}$$

it follows that the *active input admittance* Y_{mn} of the mnth element in the array can be expressed in terms of the mutual admittances and the exciting voltages by

$$Y_{mn} = \sum_{p=-M}^{M}\sum_{q=-N}^{N}\frac{V_{pq}}{V_{mn}}Y_{mn,pq}. \tag{3.22}$$

These relationships apply quite generally to elements with completely arbitrary pattern space factors, whether alike or not. But if an array is to possess an array factor the individual pattern space factors must be identical to each other. This is an assumption that is always implicit in the very concept of array factor. Although never realized exactly in practice it is an assumption that can be approximated very closely under certain circumstances. Thus, for an array factor to exist it is necessary to assume that all of the individual space factors (3.17) can be approximated by some single function $P_x(k_x, k_y, k)$,

$$P_{xmn}(k_x, k_y, k) \equiv P_x(k_x, k_y, k). \tag{3.23}$$

With this assumption the active input admittance Y_{mn} then becomes

$$Y_{mn} = \frac{1}{(2\pi)^2 kZ_0}\int_{-\infty}^{\infty}\int_{-\infty}^{\infty}\frac{(k^2-k_y^2)\,|P_x(k_x, k_y, k)|^2}{(k^2-k_x^2-k_y^2)^{\frac{1}{2}}}\times$$

$$\times\left\{\frac{\sum\limits_{p=-M}^{M}\sum\limits_{q=-N}^{N}V_{pq}\,e^{i(pk_xd_x+qk_yd_y)}}{V_{mn}\,e^{i(mk_xd+nk_yd_y)}}\right\}dk_x\,dk_y. \tag{3.24}$$

The double summation in the bracketed part of the integrand is the ordinary array factor for isotropic elements, whereas the unbracketed part is proportional to the conjugate of the complex power density of a single element. Thus it is concluded that *the effect of the array factor on the input admittance of any particular element in an array is to act as a complex weighting factor on the complex power density of that element in the* (k_x, k_y) *plane.*

An important feature of this expression for input admittance is that it admits of a simple geometric interpretation for the interaction between the pattern space factor of the individual elements and the array factor of the lattice configuration. In the case of an array of narrow half-wave slots with identical sinusoidal current distributions, for example, the unbracketed part of the integrand of (3.24) is proportional to the conjugate of the complex power density plotted in Fig. 3.1 for each element alone. The bracketed part then shows how the power density plot of the *mn*th element will be modified by the presence of all of the other elements in the array. Thus, the input admittance of an element in an array is represented geometrically by the volume under a surface whose ordinate is proportional to power density, in much the same way as for a single isolated element. This is simply a more general case of the analytic extension of the Poynting vector method.

Arraying can have a peculiar effect, however, upon the origin of the radiative and reactive contributions in the (k_x, k_y) plane. For a single isolated element the radiative contributions all arise from within the visible region alone and the reactive contributions all arise from within the invisible region alone. This is evident from the fact that the only complex factor in the expression for power density of an isolated element is $(k^2-k_x^2-k_y^2)^{\frac{1}{2}}$, which is always purely real in the visible region and purely imaginary in the invisible region. But when that element is placed in an array the bracketed expression by which its power density is multiplied need not always be real. If the bracketed expression has an imaginary part in the visible region then that region will create a reactive contribution. Likewise an imaginary part in the invisible region will create a radiative contribution. Thus, although identification of the visible region with resistance and of the invisible region with reactance is always valid for the array as a whole, it is not always valid for an individual element in the array.

The simplest array possible is the degenerate case of a single element, occurring for $M = N = 0$. The input admittance (3.24) then reduces to just the ratio of the conjugate of the complex power (3.4) to the mean-squared input voltage $\frac{1}{2}\,|V_{00}|^2$, which is half (because of power flow on one side only) of the value obtained previously by the analytic extension of the Poynting vector method for a single element radiating equally on both sides of the aperture.

The next most simple array is one with two elements. Borgiotti† has obtained an expression similar to (3.19) for the mutual admittance between two identical elements and has shown that the numerical results obtained for thin half-wave slots of various orientations are in essential agreement with earlier results obtained by P. S. Carter using the induced-e.m.f. method.

† G. V. Borgiotti. A novel expression for the mutual admittance of planar radiating elements. *IEEE Trans. Antennas Propagat.* **AP-16,** pp. 329–33, May 1968.

Any array whose main beam can be scanned electronically by controlling the relative phase of the excitation of its elements is called a *phased array*. The simplest analytical description for a phased array is obtained when the individual elements are all excited with equal magnitudes V_0 and with a uniform phase progression such that the main beam lies in an arbitrarily prescribed direction (k_{x0}, k_{y0}),

$$V_{mn} = V_0 e^{-i(mk_{x0}d_x + nk_{y0}d_y)}. \tag{3.25}$$

The direction of the main beam can be scanned electronically by changing the values of the phasing constants k_{x0} and k_{y0}. For such an array the array factor then separates into a product of independent functions of k_x and k_y, reducing to

$$\sum_{p=-M}^{M} \sum_{q=-N}^{N} V_{pq} e^{i(pk_x d_x + qk_y d_y)} =$$

$$= V_0 \left\{ \sum_{p=-M}^{M} e^{ip(k_x - k_{x0})d_x} \right\} \left\{ \sum_{q=-N}^{N} e^{iq(k_y - k_{y0})d_y} \right\} =$$

$$= V_0 \frac{\sin(M+\frac{1}{2})(k_x - k_{x0})d_x}{\sin \frac{1}{2}(k_x - k_{x0})d_x} \times \frac{\sin(N+\frac{1}{2})(k_y - k_{y0})d_y}{\sin \frac{1}{2}(k_y - k_{y0})d_y}. \tag{3.26}$$

This particular dependence on k_x and k_y is identical to that of diffraction by optical gratings.† It is a periodic function of k_x and k_y separately, with *grating lobes* appearing at intervals of $2\pi/d_x$ and $2\pi/d_y$, respectively. For increasingly large values of M and N it becomes concentrated more and more about the grating lobes, rising to the value $(2M+1)(2N+1)$ at the peak of each lobe and falling rapidly to relatively small values elsewhere. Each of the two factors thus approaches a function having an infinite sequence of equally spaced delta-like singularities. The strength of the singularities can be determined by integrating over a single period. For the grating-lobe function of k_x the strength of the singularities is

$$\int_{-\frac{\pi}{d_x}}^{\frac{\pi}{d_x}} \frac{\sin(M+\frac{1}{2})(k_x - k_{x0})d_x}{\sin \frac{1}{2}(k_x - k_{x0})d_x} \, d(k_x - k_{x0}) = \frac{2\pi}{d_x}. \tag{3.27}$$

In the limit as M approaches infinity the grating-lobe function of k_x then approaches an infinite series of delta functions,

$$\frac{\sin(M+\frac{1}{2})(k_x - k_{x0})d_x}{\sin \frac{1}{2}(k_x - k_{x0})d_x} \sim \frac{2\pi}{d_x} \sum_{r=-\infty}^{\infty} \delta\{(k_x - k_{x0})d_x - 2\pi r\}, \tag{3.28}$$

† M. Born and E. Wolf. *Principles of optics*, Pergamon Press, New York, 1959, p. 403.

which is independent of the value of M itself. Similar considerations apply for the grating-lobe function of k_y in the limit as N approaches infinity.

With this limiting functional form for the array factor the input admittance (3.24) of a single element in an infinitely large array then reduces simply to

$$Y_{mn} \sim \frac{1}{kZ_0 d_x d_y} \sum_{r=-\infty}^{\infty} \sum_{s=-\infty}^{\infty} \left\{ \frac{(k^2-k_y^2)\,|P_x(k_x, k_y, k)|^2}{(k^2-k_x^2-k_y^2)^{\frac{1}{2}}} \bigg|_{\substack{k_x=k_{x0}+(2\pi r/d_x)\\ k_y=k_{y0}+(2\pi s/d_y)}} \right\}. \quad (3.29)$$

It is independent of the location indices m and n of the element in the array,

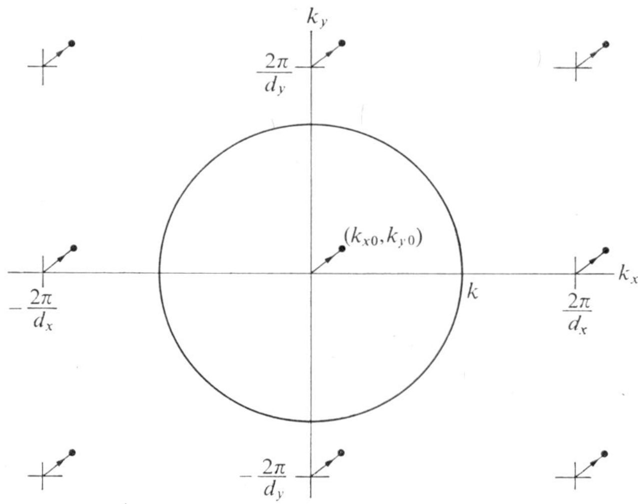

FIG. 3.4. Grating-lobe sampling of the power density plot of a single element in an infinite array. When the main beam is shifted from the broadside direction (0, 0) to the direction (k_{x0}, k_{y0}) the entire grating-lobe pattern shifts with it as indicated.

as it must be for a truly infinite array. This is a special case of a result obtained independently by Borgiotti.†

The series (3.29) has an intuitively appealing geometric interpretation. Each term is proportional to a sample value in the (k_x, k_y) plane of the conjugate of the power density of a single element. The sampling points are at the grating lobes of the array factor. When the main beam is in the broadside direction they are at $k_x = 2\pi r/d_x$, $k_y = 2\pi s/d_y$. The effect of scanning is to shift the entire pattern of grating lobes by a distance k_{x0} in the k_x direction and k_{y0} in the k_y direction, as indicated in Fig. 3.4. All of the grating lobes that lie within the visible region $k_x^2+k_y^2 < k^2$ contribute exclusively to radiation conductance alone, while all those that lie outside contribute exclusively to input susceptance alone. Thus, the

† G. V. Borgiotti (1968), *loc. cit.*, eqn (23).

input admittance of any element in an infinite array is determined in part by the power density plot of a single element and in part by the spacing chosen between the elements in the lattice configuration. By visualizing the way in which these two properties interact one can distinguish clearly between the effects that are caused by the design of the element and those that are caused by the design of the lattice configuration. A variety of element designs and lattice configurations has been investigated by Diamond† and by Ehlenberger *et al.*‡

An interesting, but less appealing, geometric interpretation of (3.29) has been proposed by Wheeler.§ It is obtained by rewriting (3.29) in the following form,

$$Y_{mn} \sim \left[\frac{1}{kZ_0 d_x d_y}\sum_{r=-\infty}^{\infty}\sum_{s=-\infty}^{\infty} \times \frac{\left\{k^2-\left(k_y+\dfrac{2\pi s}{d_y}\right)^2\right\}\left|P_x\left(k_x+\dfrac{2\pi r}{d_x},k_y+\dfrac{2\pi s}{d_y},k\right)\right|^2}{\left\{k^2-\left(k_x+\dfrac{2\pi r}{d_x}\right)^2-\left(k_y+\dfrac{2\pi s}{d_y}\right)^2\right\}^{\frac{1}{2}}}\right]_{\substack{k_x=k_{x0}\\k_y=k_{y0}}}. \quad (3.30)$$

In this form it can be interpreted as being proportional to just a single sample taken from the sum of an infinite number of identical power density plots, in contrast to (3.29) which was proportional to the sum of an infinite number of samples taken from a single power density plot.

Wheeler's geometric interpretation of (3.30) is illustrated in Fig. 3.5. Identical power density plots of the element pattern are placed on the (k_x, k_y) plane with their centres located periodically at the points $k_x = 2\pi r/d_x$, $k_y = 2\pi s/d_y$. The input admittance of any element in an infinite array is then proportional to the conjugate of the sum of the values obtained from all of these power density plots at just the single sampling point $k_x = k_{x0}$, $k_y = k_{y0}$. The value obtained from the inside of any given circle contributes to radiation conductance only, while the value obtained from the outside of any given circle contributes to input susceptance only, in exactly the same way as in Fig. 3.4. Wheeler refers to these power density plots, after dividing

† B. L. Diamond. A generalized approach to the analysis of infinite planar array antennas. *Proc. IEEE*, **56,** pp. 1837–51, Nov. 1968.

‡ A. G. Ehlenberger, L. Schwartzman, and L. Topper. Design criteria for linearly polarized waveguide arrays. *Proc. IEEE*, **56,** pp. 1861–72, Nov. 1968.

§ H. A. Wheeler. The grating lobe series for the impedance variation in a planar phased-array antenna. *IEEE Trans. Antennas Propagat.* **AP-13,** pp. 825–7, Sept. 1965; also **AP-14,** pp. 707–14, Nov. 1966.

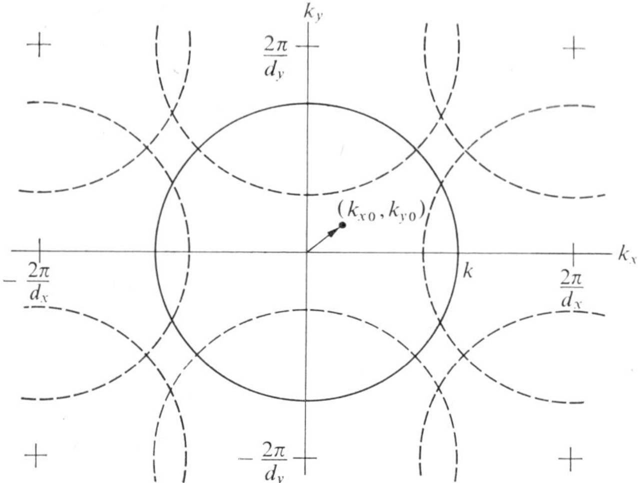

FIG. 3.5. Wheeler's geometric interpretation of grating-lobe sampling of the power density plot of a single element in an infinite array.

by the mean-squared value of the input current,† as 'impedance craters' (because of their similarity in appearance to geological craters) and to the sum (3.30) as the 'grating lobe series of impedance craters'.

The reason why Wheeler's geometric interpretation of his grating lobe series (3.30) was said to be less appealing than the geometric interpretation of (3.29) is that it cannot be extended to the case of finite arrays. In fact, it has no meaning at all for finite arrays, whereas the geometric interpretation of (3.29) arose directly from the limiting case of finite arrays. The effect upon Fig. 3.4 that would be caused by truncating an infinite array to some finite size would be simply a broadening of the grating lobes. But the effect upon Fig. 3.5 that would be caused by truncation cannot even be discussed because the figure itself would become meaningless.

3.4. Meaning of quality factor Q

All physically realizable solutions to antenna synthesis problems must be constrained, either explicitly or implicitly, in such a way as to ensure that the input bandwidth will be wide enough to pass the full information content of the signals to be transmitted. As an alternative to bandwidth it is widely presumed that one can resort to a simpler quantity, the quality factor Q, whose reciprocal approaches the bandwidth as the bandwidth approaches zero. Contrary to popular opinion, however, the usual definition of Q in terms of stored magnetic and electric energies will be seen to have no connection whatsoever with bandwidth in the case of radiating systems. In fact,

† Wheeler worked with impedances instead of admittances.

it is quite possible that its presumed relationship to bandwidth may not be valid for any electromagnetic system other than the special case of series or parallel RLC circuits from which the concept of Q as a measure of input bandwidth first arose.

For all highly resonant electromagnetic systems, of which series or parallel RLC circuits are but special cases, the relative input bandwidth B is known to approach zero asymptotically as

$$B \sim \left. \frac{2\langle P \rangle}{\omega \dfrac{d}{d\omega}\{2\omega(\langle W_m \rangle - \langle W_e \rangle)\}} \right|_{\omega=\omega_r} \tag{3.31}$$

in terms of the time-average magnetic and electric energies stored and power dissipated, $\langle W_m \rangle$, $\langle W_e \rangle$, and $\langle P \rangle$, respectively, at the resonance frequency ω_r. Resonance occurs when the time-average magnetic and electric energies are equal. In the particular case of series or parallel RLC circuits at resonance, one finds that the frequency derivative of ω times the difference between the time-average magnetic and electric energies is exactly equal to just the total time-average energy stored in the circuit and, hence, that this asymptotic value of relative bandwidth approaches the reciprocal of the conventional definition for quality factor Q,

$$Q = \left. \frac{\omega(\langle W_m \rangle + \langle W_e \rangle)}{\langle P \rangle} \right|_{\omega=\omega_r}. \tag{3.32}$$

The resonance condition $\omega = \omega_r$ is essential, for without it there would be no basis at all for expecting that Q is related to B. If this asymptotic reciprocal relationship between B and Q were to continue to hold for all electromagnetic systems, as is usually assumed, then the parameter B, which is measurable, could always be approximated by the simpler but nonmeasurable parameter Q. The reason why it is simpler is that the calculation of Q does not require that the frequency dependence of the system be known (Q depends upon the stored energies at only a single frequency ω_r), whereas the calculation of B does. But there is no valid reason to believe that this asymptotic reciprocal relationship should apply to any system other than the series or parallel RLC circuits from which it first arose. In the case of all radiating systems, in particular, it is known to fail completely because of the fact that the total time-average energy stored in such systems is always infinite. Such failures do not necessarily mean that the concept of Q as an approximate measure of input bandwidth (3.31) must be abandoned, however. On the contrary, if Q is defined properly it will be shown here to lead to results that are fully compatible with (3.31).

The crux of the difficulty lies in a fundamental logical inconsistency that is implicit in the classical definition (3.32) for Q. The bandwidth (3.31) that the reciprocal of Q is intended to represent depends not upon the time-average stored energies $\langle W_m \rangle$ and $\langle W_e \rangle$ individually but only upon their difference. The logical conclusion, then, is that only those parts of $\langle W_m \rangle$ and $\langle W_e \rangle$ that contribute to that difference can have any observable effect upon the bandwidth measured and, hence, should be the only parts to appear in the definition of Q. The classical definition (3.32), on the other hand, is not limited to just the observable parts alone but depends, instead, upon the whole of $\langle W_m \rangle$ and $\langle W_e \rangle$. Thus, any parts of $\langle W_m \rangle$ and $\langle W_e \rangle$ that are identically equal at all frequencies ω, and hence that could not in any way affect the value of the bandwidth (3.31), would still have a very definite effect upon the value of (3.32). It is concluded, therefore, that there is no valid basis for the usual presumption that the reciprocal of the conventional value of Q as defined by (3.32) must approach the value of the bandwidth B.

This long-standing logical inconsistency can be removed by expressing Q in terms of just the part $\langle\langle W_m \rangle\rangle$ of $\langle W_m \rangle$, and the part $\langle\langle W_e \rangle\rangle$ of $\langle W_e \rangle$, that has an observable effect upon B. Making this fundamental change in the definition of Q its value becomes

$$Q = \frac{\omega(\langle\langle W_m \rangle\rangle + \langle\langle W_e \rangle\rangle)}{\langle P \rangle}\bigg|_{\omega=\omega_r} \tag{3.33}$$

in terms of the *observable stored energies* $\langle\langle W_m \rangle\rangle$ and $\langle\langle W_e \rangle\rangle$. It is now internally consistent with bandwidth because it is determined by precisely the same quantities that determine bandwidth.

The determination of Q thus depends upon being able to determine those parts of the time-average magnetic and electric stored energies that have an observable effect upon the relative input bandwidth. Exact formulas for the observable stored energies in the fields of planar sources are derived in the next section.

3.5. Observable stored energies of planar sources[4]

The origin of the concept of time-average stored energies that appear in the asymptotic expression (3.31) for input bandwidth is the volume integrals in the complex Poynting theorem. For a planar source over an aperture A that is radiating into a volume V of free space enclosed by the aperture plane $z = 0$ and a surface S in the half-space $z > 0$, the complex Poynting theorem is given by

$$-\tfrac{1}{2}\int_A \mathbf{E}\times\mathbf{H}^*.\mathbf{n}\,da = +\tfrac{1}{2}\int_S \mathbf{E}\times\mathbf{H}^*.\mathbf{n}\,da + i2\omega\left(\frac{\mu_0}{4}\int_V \mathbf{H}.\mathbf{H}^*\,dv - \frac{\epsilon_0}{4}\int_V \mathbf{E}.\mathbf{E}^*\,dv\right). \tag{3.34}$$

It comes from the system equations (2.3), and is simply a mathematical identity that is satisfied by all field functions $\mathbf{E}(x, y, z, \omega)$ and $\mathbf{H}(x, y, z, \omega)$ satisfying (2.3). Its physical significance arises from the fact that the real and imaginary parts of the left-hand side represent the time-average radiative and reactive power flow, respectively, that would be produced across the aperture by a sinusoidal signal of real frequency ω at the input to the antenna. If the dimensions of the surface S are unbounded then the surface integral on the right-hand side of (3.34) will become purely real and, hence, equal to the radiated power across the aperture. Consequently the reactive power across the aperture must be equal to 2ω times the difference between the volume integrals of $(\mu_0/4)\mathbf{H}.\mathbf{H}^*$ and $(\epsilon_0/4)\mathbf{E}.\mathbf{E}^*$ over the entire half-space into which the source radiates. These volume integrals have traditionally been interpreted as representing time-average magnetic and electric energy stored within the volume, based on the hypothesis that the formal expressions for magnetostatic and electrostatic energy storage carry over to the case of time-harmonic electromagnetic fields. For an infinite volume these volume integrals always become infinite, which means that *the time-average stored energies in radiating systems are always infinite.* But it is important to note that the physical interpretation of the complex Poynting theorem itself, when viewed as an expression for complex power flow through an aperture, says nothing at all about the interpretation of the two volume integrals individually, only about their *difference.* This is an essential point in arriving at an internally consistent representation for the observable stored magnetic and electric energies individually.

The volume integrals in (3.34) will now be evaluated for an arbitrary planar source in terms of its pattern space factor by treating the infinite half-space $z > 0$ as the limit of the interior of a rectangular parallelopiped bounded by the planes $x = \pm x_0$, $y = \pm y_0$, $z = z_0$ (Fig. 3.6) when x_0, y_0, and z_0 approach infinity:

$$\frac{\mu_0}{4}\int_V \mathbf{H}.\mathbf{H}^*\,dv = \frac{\mu_0}{4}\lim_{x_0,y_0,z_0\to\infty}\int_{-x_0}^{x_0}\int_{-y_0}^{y_0}\int_0^{z_0}(H_xH_x^*+H_yH_y^*+H_zH_z^*)\,dx\,dy\,dz, \tag{3.35a}$$

$$\frac{\epsilon_0}{4}\int_V \mathbf{E}.\mathbf{E}^*\,dv = \frac{\epsilon_0}{4}\lim_{x_0,y_0,z_0\to\infty}\int_{-x_0}^{x_0}\int_{-y_0}^{y_0}\int_0^{z_0}(E_xE_x^*+E_yE_y^*+E_zE_z^*)\,dx\,dy\,dz. \tag{3.35b}$$

Upon substituting the field expressions (2.16) and interchanging the order of integration the triple integral on x, y, z approaches the following limit as x_0

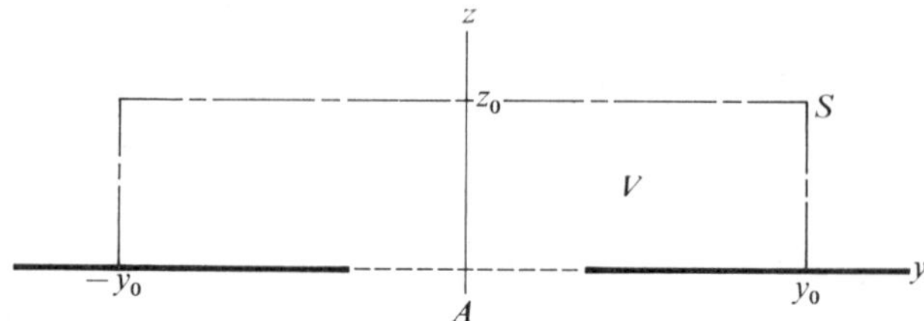

FIG. 3.6. A rectangular volume V contained within a closed surface bounded by S and the aperture plane $z = 0$.

and y_0 become infinite:

$$\int_{-\infty}^{\infty}\int_{-\infty}^{\infty}\int_{0}^{z_0}\exp[-i\{(k_x-k_x')x+(k_y-k_y')y+$$

$$+[(k^2-k_x^2-k_y^2)^{\frac{1}{2}}-\{(k^2-k_x'^2-k_y'^2)^{\frac{1}{2}}\}^*]z\}]\,dx\,dy\,dz =$$

$$=\begin{cases} z_0(2\pi)^2\,\delta(k_x-k_x')\,\delta(k_y-k_y'), & k_x^2+k_y^2<k^2 \\ \dfrac{1-\exp\{-2(k_x^2+k_y^2-k^2)^{\frac{1}{2}}z_0\}}{2(k_x^2+k_y^2-k^2)^{\frac{1}{2}}}(2\pi)^2\delta(k_x-k_x')\,\delta(k_y-k_y'), & k_x^2+k_y^2>k^2. \end{cases}$$

(3.36)

The volume integrals (3.35) of the field functions then become integrals of the pattern space factor over the wavenumber plane,

$$\frac{\mu_0}{4}\int_V \mathbf{H}.\mathbf{H}^*\,dv = \frac{1}{2\omega(2\pi)^2 2kZ_0}\lim_{z_0\to\infty}\Bigg[\iint_{k_x^2+k_y^2<k^2}\Bigg\{\frac{|k_xk_yF_x+(k^2-k_x^2)F_y|^2}{k^2-k_x^2-k_y^2}+$$

$$+\frac{|(k^2-k_y^2)F_x+k_xk_yF_y|^2}{k^2-k_x^2-k_y^2}+|k_yF_x-k_xF_y|^2\Bigg\}z_0\,dk_x\,dk_y+$$

$$+\iint_{k_x^2+k_y^2>k^2}\Bigg\{\frac{|k_xk_yF_x+(k^2-k_x^2)F_y|^2}{k_x^2+k_y^2-k^2}+$$

$$+\frac{|(k^2-k_y^2)F_x+k_xk_yF_y|^2}{k_x^2+k_y^2-k^2}+|k_yF_x-k_xF_y|^2\Bigg\}\times$$

$$\times\frac{1-\exp\{-2(k_x^2+k_y^2-k^2)^{\frac{1}{2}}z_0\}}{2(k_x^2+k_y^2-k^2)^{\frac{1}{2}}}\,dk_x\,dk_y\Bigg], \qquad (3.37a)$$

$$\frac{\epsilon_0}{4}\int_V \mathbf{E}.\mathbf{E}^*\,dv = \frac{1}{2\omega(2\pi)^2 2kZ_0}\lim_{z_0\to\infty}\left[\iint_{k_x^2+k_y^2<k^2} k^2\left\{|F_x|^2+|F_y|^2+\right.\right.$$

$$\left.+\frac{|k_xF_x+k_yF_y|^2}{k^2-k_x^2-k_y^2}\right\}z_0\,dk_x\,dk_y + \iint_{k_x^2+k_y^2>k^2} k^2\left\{|F_x|^2+|F_y|^2+\right.$$

$$\left.\left.+\frac{|k_xF_x+k_yF_y|^2}{k_x^2+k_y^2-k^2}\right\}\frac{1-\exp\{-2(k_x^2+k_y^2-k^2)^{\frac{1}{2}}z_0\}}{2(k_x^2+k_y^2-k^2)^{\frac{1}{2}}}\,dk_x\,dk_y\right]. \tag{3.37b}$$

All four of the wavenumber integrals are infinite for arbitrary aperture-limited functions $F_x(k_x, k_y, k)$ and $F_y(k_x, k_y, k)$ because each integral has a logarithmic singularity at the circle $k_x^2+k_y^2 = k^2$. Consequently *neither of the two volume integrals* (3.37) *exists individually*. The singular points correspond physically to uniform plane waves moving parallel to the aperture plane at the velocity of light. They are points at one end of a continuum of phase velocities that extends above the velocity of light (visible, uniform plane waves) and also below it (invisible, nonuniform plane waves), thereby causing the wavenumber integrals over the visible and invisible regions individually to have an infinite limit at those end points.

Upon detailed comparison of the two wavenumber integrals over the visible region $k_x^2+k_y^2 < k^2$ it will be seen that their integrands are identically equal to one another, for any choice of the frequency and of the pattern space factor, and hence that their difference vanishes identically. The visible region, then, makes no contribution whatsoever to the observable stored energies. All of the reactive effects must come solely from the invisible region.

Consider first the reactive energies stored by the three field components $H_z(x, y, z, k)$, $E_x(x, y, z, k)$, and $E_y(x, y, z, k)$. Each is represented by its wavenumber integral over just the invisible region $k_x^2+k_y^2 > k^2$ alone, which is always finite since it has no singularity at the circle $k_x^2+k_y^2 = k^2$. Hence, each must be the complete and unique time-average reactive energy stored by that particular field component.

Consider next the contribution from the invisible region by the other three field components $H_x(x, y, z, k)$, $H_y(x, y, z, k)$, and $E_z(x, y, z, k)$. Each is infinite. Physically, then, each would be expected to be unobservable because the infinities do not appear in the expression (3.4) for complex power at the aperture. But that is only partly true. Although the infinities are, indeed, unobservable because they cancel identically, the cancellation must leave certain finite residues necessary to account for (3.4). The residues do not represent the actual reactive energies stored by each of these components individually because a large part of the contribution by each component is missing: the infinities have been cancelled. But these residues can be expressed in terms of quantities already known to represent reactive stored energies of the proper type. The residue of magnetic energy can be expressed in terms of the reactive energy stored by just the magnetic field component $H_z(x, y, z, k)$, and the residue of electric energy can be expressed in terms of the reactive energy stored by just the electric field components $E_x(x, y, z, k)$ and $E_y(x, y, z, k)$. This follows from the observation that the integrands of both of the wavenumber integrals over the invisible region would have been identically equal and their difference would have vanished identically, just as in the visible region, if the sign of their nonsingular terms had been negative instead of positive. Hence, the infinities arising from the three singular terms always cancel identically except for finite residues that are equal in value to the reactive energies stored by $H_z(x, y, z, k)$, $E_x(x, y, z, k)$, and $E_y(x, y, z, k)$.

Thus, the only parts of the volume integrals (3.37) that have a physically observable effect upon the input bandwidth and that also can be expressed in terms of actual reactive stored energies of the proper type is twice the reactive energies stored by $H_z(x, y, z, k)$, $E_x(x, y, z, k)$, and $E_y(x, y, z, k)$ (twice because the values of the residues left from the infinite energies stored by $H_x(x, y, z, k)$, $H_y(x, y, z, k)$, and $E_z(x, y, z, k)$ are exactly equal to the actual reactive energies stored by $H_z(x, y, z, k)$, $E_x(x, y, z, k)$, and $E_y(x, y, z, k)$). When these observable stored energies are multiplied by 2ω they represent individually the observable inductive and capacitive powers supplied to the infinite volume by the radiating source:

$$2\omega\langle\langle W_m\rangle\rangle_A = \frac{1}{(2\pi)^2 2kZ_0}\iint\limits_{k_x^2+k_y^2>k^2}\frac{|k_yF_x-k_xF_y|^2}{(k_x^2+k_y^2-k^2)^{\frac{1}{2}}}\,dk_x\,dk_y, \tag{3.38a}$$

$$2\omega\langle\langle W_e\rangle\rangle_A = \frac{1}{(2\pi)^2 2kZ_0}\iint\limits_{k_x^2+k_y^2>k^2}\frac{k^2(|F_x|^2+|F_y|^2)}{(k_x^2+k_y^2-k^2)^{\frac{1}{2}}}\,dk_x\,dk_y. \tag{3.38b}$$

They are always nonnegative, and are always convergent at the circle $k_x^2+k_y^2 = k^2$ because of the fact that $F_x(k_x, k_y, k)$ and $F_y(k_x, k_y, k)$ are entire analytic functions of k_x and k_y. Furthermore their difference is identical, term for term, to the net reactive power (3.6) that was obtained by integrating the complex Poynting vector over the radiating aperture.

The sum of these observable reactive powers gives the total observable reactive power at the aperture,

$$2\omega(\langle\langle W_m\rangle\rangle_A+\langle\langle W_e\rangle\rangle_A) = \frac{1}{(2\pi)^2 2kZ_0}\times$$

$$\times\iint\limits_{k_x^2+k_y^2>k^2}\frac{(k_y^2+k^2)|F_x|^2+(k_x^2+k^2)\,|F_y|^2-2k_xk_y\,\mathrm{Re}\,F_xF_y^*}{(k_x^2+k_y^2-k^2)^{\frac{1}{2}}}\,dk_x\,dk_y. \tag{3.39}$$

It is always finite, which means that the value of Q obtained from the new definition (3.33) will also be finite. Although this formula (3.39) for *total* observable reactive power and formula (3.6) for *net* reactive power are strikingly similar in appearance they have quite different consequences: the total observable reactive power is always positive and can be extremely large, as in the case of certain superdirective sources, and yet the net reactive power can be zero (resonance) or even negative.

In the case of complementary apertures the fields $\mathbf{E}(x, y, z, k)$ and $Z_0\mathbf{H}(x, y, z, k)$ interchange, according to (2.47), and the pattern space factor components $F_x(k_x, k_y, k)$ and $F_y(k_x, k_y, k)$ are obtained from $Z_0H_x(x, y, z, k)$ and $Z_0H_y(x, y, z, k)$, respectively, instead of from $E_x(x, y, z, k)$ and $E_y(x, y, z, k)$. Consequently expressions (3.38*a*) and (3.38*b*) for the observable reactive powers simply interchange.

An alternative representation for the observable reactive powers (3.38) that is sometimes more useful is obtained by representing the pattern space factor in terms of its polar components $F_\kappa(\kappa, \phi, k)$ and $F_\phi(\kappa, \phi, k)$. After making the transformation from rectangular to polar components it becomes

$$2\omega\langle\langle W_m\rangle\rangle_A = \frac{1}{(2\pi)^2 2kZ_0}\int\limits_k^\infty\left[\int\limits_{-\pi}^{\pi}|F_\phi(\kappa, \phi, k)|^2\,d\phi\right]\frac{\kappa^3\,d\kappa}{(\kappa^2-k^2)^{\frac{1}{2}}}, \tag{3.40a}$$

$$2\omega\langle\langle W_e\rangle\rangle_A = \frac{k^2}{(2\pi)^2 2kZ_0}\int\limits_k^\infty\left[\int\limits_{-\pi}^{\pi}\{|F_\kappa(\kappa, \phi, k)|^2+|F_\phi(\kappa, \phi, k)|^2\}\,d\phi\right]\frac{\kappa\,d\kappa}{(\kappa^2-k^2)^{\frac{1}{2}}}. \tag{3.40b}$$

In this form it is seen that the observable magnetic energy, determined by just the angular component $F_\phi(\kappa, \phi, k)$ alone, will be zero whenever $F_\phi(\kappa, \phi, k)$ is identically zero. Thus, in the special case of circularly symmetric radial sources that will be treated in section 4.4 the total observable stored energy will be purely electric.

All of the conclusions reached here on the observable stored energies of planar sources can be summarized quite concisely by noting first that the nonsingular terms in the integrands of the wavenumber integrals in (3.37) arise from the rectangular field components that vanish on the plane outside of the aperture, whereas the singular terms that create the infinities arise from the rectangular field components that terminate on electric currents or charges on the conducting plane. The same is true for complementary apertures in terms of fictitious magnetic currents or charges that would exist on a magnetic conducting plane. In either case the observable magnetic energy is twice that produced in the invisible region of the pattern space factor by the magnetic field components that vanish in the plane outside of the aperture, while the observable electric energy is twice that produced in the invisible region by the electric field components that vanish in the plane outside of the aperture. Thus, it is concluded quite generally that *the only physically observable part of the volume integrals usually assumed to represent time-average stored energies in the fields of any planar source is twice that produced in the invisible region of the pattern space factor by those field components that vanish in the aperture plane outside of the source.*

Perhaps it should be stated explicitly that there is no inconsistency in the conclusion that the observable stored energies come from the invisible region, despite any possible semantical implications to the contrary that might be suggested by the terms 'observable' and 'invisible'. The term 'observable', as used here, refers to the fact that these are the only parts of the time-average stored energies that have a measurable effect at the input port. The term 'invisible', on the other hand, refers to the quite different fact that the part of the pattern space factor which determines these observable stored energies cannot be measured in the far field; i.e. it is not identifiable with the radiation pattern.

The concept of observable stored energies is admittedly a rather radical departure from classical electromagnetic theory, although one that appears to be absolutely essential if the concept of quality factor as an approximate measure for relative input bandwidth is to have any meaning at all. No universal formulation for the concept is known as yet, nor does the concept itself appear to have gained general acceptance. Even in the special case of planar sources the validity of formulas (3.38) has been contested on grounds of uniqueness.† The only known test of these formulas is for the case of planar dipole antennas treated in section 3.7, where it will be shown that the value of Q obtained from an assumed sinusoidal current distribution approaches a value that is close to the reciprocal of the bandwidth.

3.6. Causal behaviour of filamentary dipole antennas

One of the obstacles to the establishment of a relationship between bandwidth and Q was partially overcome in section 3.4 by the recognition that Q must be described in terms of those parts of the stored magnetic and electric energies that are observable in the sense that they contribute to the bandwidth seen at the input port. In the case of idealized slot or dipole antennas that particular obstacle was overcome completely in section 3.5 by the development of explicit formulas for the observable stored energies of planar sources. But still another obstacle exists: the determination of input bandwidth requires that the frequency dependence of the system functions be known and that it be causal. By a causal system function is meant that the impulse response at any distance r from the input port must be zero prior to the time

† G. V. Borgiotti. On the reactive energy of an aperture; R. E. Collin. Stored energy, Q, and frequency sensitivity of planar aperture antennas; *IEEE Trans. Antennas Propagat.*, **AP-15**, pp. 565–9, July 1967.

r/c after the occurrence of the excitation impulse at the input port. It is this physical requirement that leads to analyticity of the system functions in the lower half of the complex k-plane at any distance r from the input port.

The only radiating system whose frequency dependence is known exactly is the filamentary dipole antenna; i.e. the limiting case of a dipole antenna of zero width. The current distribution along a dipole antenna of arbitrary length is known, experimentally as well as theoretically, to approach a sinusoid as the width approaches zero. It will now be shown that the impulse response everywhere along the length of such a filamentary dipole antenna is truly causal, in spite of the somewhat surprising fact that the sinusoidal current distribution (3.7) from which it will be obtained is not quite correct physically because it is not analytic in the lower half of the complex k-plane; it has an essential singularity at $k = -i\infty$.

The impulse response of current density anywhere on a planar dipole antenna of arbitrary length b and vanishing width a, when obtained from the Fourier transform of the sinusoidal current density (3.7), is found to be

$$\hat{\mathscr{K}}_y(x, y, \tau) = \frac{1}{2\pi}\int_{-\infty}^{\infty} K_y(x, y, \omega/c)e^{i\omega\tau}\,d\omega$$

$$= \frac{I_{\max}}{i4a}\left\{\delta\left(\tau+\frac{\frac{1}{2}b-|y|}{c}\right)-\delta\left(\tau-\frac{\frac{1}{2}b-|y|}{c}\right)\right\}. \tag{3.41}$$

Each of the two delta functions can be interpreted physically as representing a pair of impulses of electric current density, one at $+y$ and the other at $-y$, travelling in opposite directions along the two halves of the wire at the velocity c of waves in free space. Usually the time-reference $\tau = 0$ is supposed to denote the moment at which the response to the excitation impulse first appears at the input port. Here, however, the moment at which the response first appears at the input port is not $\tau = 0$ but $\tau = -b/2c$, as seen from the first of the two delta functions when evaluated at the input port $y = 0$. The response represented by the first delta function thus consists of a pair of impulses that move outward from the input port at time $\tau = -b/2c$ with velocity c along the two halves of the wire, without attenuation (i.e. there is no radiation in the limiting case of a filamentary dipole antenna), until they reach the two ends $y = \pm b/2$ at time $\tau = 0$. At that moment a second pair of impulses appears, represented by the second of the two delta functions, whose strengths are equal to those of the outward moving pair but which are moving inward toward the input port. This incoming pair of impulses can be interpreted physically as the reflection of the outgoing pair from the two ends of the wire. The negative sign before the second delta function represents a reversal of phase upon reflection, giving total cancellation of the two at the ends of the wire at time $\tau = 0$. The resulting sequence of events, in which a pair of impulses originates at the input port at time $\tau = -b/2c$ and then returns again at time $\tau = +b/2c$ after reflection from the two ends, is clearly a complete physical description of a true causal process, even though the time reference has been shifted from $\tau = 0$ to $\tau = -b/2c$.

This shift in the time reference is responsible for the fact that the sinusoidal current distribution (3.7) is not analytic in the lower half of the complex k-plane. It can be changed into a causally correct analytic function simply by moving the time reference ahead from $\tau = -b/2c$ to $\tau = 0$. Thus, by making a change of variable $\tau' = \tau + (b/2c)$ in the representation for impulse response the causally correct current distribution is seen to be

$$\int_{-\infty}^{\infty} \hat{\mathscr{K}}_y\{x, y, \tau' - (b/2c)\} e^{-i\omega\tau'}\, d\tau'$$

$$= \frac{I_{\max}}{i4a}\left[\exp\left\{-i\omega\left(\frac{b}{2c} - \frac{\frac{1}{2}b - |y|}{c}\right)\right\} - \exp\left\{-i\omega\left(\frac{b}{2c} + \frac{\frac{1}{2}b - |y|}{c}\right)\right\}\right]$$

$$= \frac{I_{\max}}{2a} e^{-i(kb/2)} \sin k(\tfrac{1}{2}b - |y|). \tag{3.42}$$

This current distribution is analytic everywhere in the lower half of the k-plane, as required, at all points on the dipole antenna. It differs from the usual expression (3.7) by only the phase factor $e^{-i(kb/2)}$, which does not affect the sinusoidal spatial dependence at all. Nor does it affect such performance characteristics as impedance, bandwidth, or Q. For that reason the usual expression (3.7) will continue to be used here to represent the current distribution of narrow planar dipole antennas.

3.7. Bandwidth and Q of narrow planar dipole antennas[4]

Two requirements must first be satisfied before any valid test can be made of the presumption that the asymptotic reciprocal relationship between bandwidth and Q holds for radiating systems: physically consistent formulas for the observable stored energies must be known in order to compute Q; and a causal description of the system must be known in order to compute bandwidth. The only antenna that satisfies both requirements at present is the narrow planar dipole or its complementary slot. Hence this is the only antenna for which such a test has been made as yet. The result of the test, which will be described below, is encouraging because it affirms that the reciprocal of the value of Q obtained from its new definition (3.33) in terms of observable stored energies does, indeed, approach the value of relative bandwidth B obtained from (3.31), at least for this particular case. It must be emphasized, however, that this affirmative conclusion applies to the one case only. There is still no evidence that it holds more generally.

Since the aperture distribution for planar dipole antennas of vanishing width is known to approach a sinusoidal function there is reason to assume that the resonance wavenumber and the observable stored energies will each approach their correct values if the distribution is taken to be sinusoidal. This assumption was made in section 3.2 to obtain the formulas (3.12) and (3.13) for radiation resistance and input reactance of vanishingly narrow dipole antennas and it will be made again to obtain formulas for bandwidth and Q.

Resonance occurs when the net reactive power (3.11) is zero or, equivalently, when the input reactance (3.13) is zero,

$$\text{Si}\,kb+\{\text{Si}\,kb-\tfrac{1}{2}\,\text{Si}\,2kb\}\cos kb+ \\ +\{\text{Cin}\,kb-\tfrac{1}{2}\,\text{Cin}\,2kb-\ln(e^{\frac{3}{2}}b/2a)\}\sin kb = 0. \quad (3.43)$$

For a given value of ka the value of kb at the first resonance is the lowest root of (3.43). The plots of input reactance shown in Fig. 3.2 indicate that at the first resonance the length of the dipole must approach half of the wavelength as the width approaches zero; i.e. the value of the lowest root $k_r b$ must approach π. The solution of (3.43) in the neighbourhood of $k_r b = \pi$, obtained by retaining only first-order terms, is found to be

$$k_r b \doteqdot \pi - \frac{\frac{1}{2}\,\text{Si}\,2\pi}{\ln(e^{\frac{3}{2}}\pi/2ka) - \text{Cin}\,\pi + \frac{1}{2}\,\text{Cin}\,2\pi}. \quad (3.44)$$

With this value for kb at resonance the value of Q and of bandwidth B can be calculated independently as follows.

Consider first the value of Q as defined by (3.33) in terms of observable stored energies. Only one of the two observable energies need be calculated because they are equal at resonance. Choose that one to be the magnetic energy (3.38*b*) (recall that the electric and magnetic energies interchange for complementary antennas),

$$2\omega\langle\langle W_m\rangle\rangle_A = \frac{k^3 Z_0\,|I_{\max}|^2}{2\pi^2}\left[\int\limits_0^k \left(\frac{\cos\frac{1}{2}k_y b - \cos\frac{1}{2}kb}{k^2-k_y^2}\right)^2 \times \right. \\ \times\left\{\int\limits_{(k^2-k_y^2)^{\frac{1}{2}}}^{\infty} \left(\frac{\sin\frac{1}{2}k_x a}{\frac{1}{2}k_x a}\right)^2 \frac{dk_x}{\{k_x^2-(k^2-k_y^2)\}^{\frac{1}{2}}}\right\} dk_y + \\ \left. +\int\limits_k^{\infty}\left(\frac{\cos\frac{1}{2}k_y b-\cos\frac{1}{2}kb}{k^2-k_y^2}\right)^2\left\{\int\limits_0^{\infty}\left(\frac{\sin\frac{1}{2}k_x a}{\frac{1}{2}k_x a}\right)^2\frac{dk_x}{\{k_x^2+(k_y^2-k^2)\}^{\frac{1}{2}}}\right\}dk_y\right] \\ \doteqdot \frac{Z_0\,|I_{\max}|^2}{2\pi^2}\int\limits_0^{\infty}\left(\frac{\cos\frac{1}{2}kb\eta - \cos\frac{1}{2}kb}{1-\eta^2}\right)^2(\tfrac{3}{2}-C-\ln\tfrac{1}{2}ka|1-\eta^2|^{\frac{1}{2}})\,d\eta. \quad (3.45)$$

An exact closed-form expression for the latter integral has been evaluated and is given by (A2.8) and (A2.9). The ratio of (3.45) to the radiated power (3.10), when evaluated at the resonance frequency (3.44), is the Q of the dipole antenna. Computed values for Q over a wide range of electrically small width are shown in Fig. 3.7.

Bandwidth (3.31) depends upon the frequency derivative of the net reactive power (3.11), the latter of which can be calculated by differentiating the

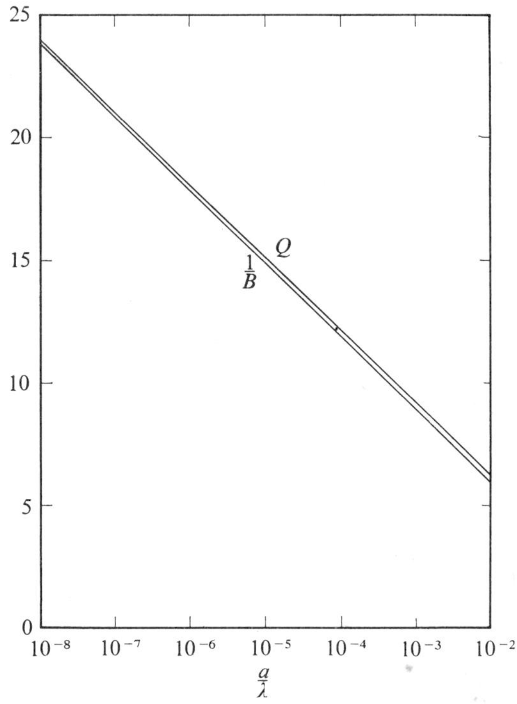

FIG. 3.7. Comparison of Q and the reciprocal of relative input bandwidth B for planar dipole antennas of small electrical width a/λ.

closed-form expression obtained from (A2.6) and (A2.7) to obtain

$$\omega \frac{d}{d\omega}\{2\omega(\langle W_m\rangle_A-\langle W_e\rangle_A)\}$$
$$\doteqdot \frac{Z_0\,|I_{\max}|^2\,kb}{8\pi}[\{(1/kb)-\text{Si}\,kb+\tfrac{1}{2}\,\text{Si}\,2kb\}\sin kb+$$
$$+\{\text{Cin}\,kb-\tfrac{1}{2}\,\text{Cin}\,2kb-\ln(e^{\frac{3}{2}}b/2a)\}\cos kb]. \quad (3.46)$$

The ratio of twice the radiative power (3.10) to this product of ω times the frequency derivative, when evaluated at resonance by (3.44), represents the relative bandwidth B of the dipole antenna. Computed values for the reciprocal of B are shown in Fig. 3.7 for comparison with the computed values of Q.

These computations of Q and $1/B$ are seen to come rather close to each other, even for widths that are not very small. It appears from the plots that they may even approach a common value as the width vanishes. This is almost, but not quite, true. The limiting form of Q as ka approaches zero is found to be

$$Q \doteqdot \frac{\pi\{\ln(2\pi e/ka)-\frac{1}{2}\,\mathrm{Cin}\,2\pi\}}{\mathrm{Cin}\,2\pi}, \tag{3.47}$$

while the limiting form of B is found to be

$$B \doteqdot \frac{\mathrm{Cin}\,2\pi}{\pi\{\ln(e^{\frac{3}{2}}\pi/2ka)-\mathrm{Cin}\,\pi+\frac{1}{2}\,\mathrm{Cin}\,2\pi\}}. \tag{3.48}$$

The limit of the difference between Q and $1/B$ is then

$$\lim_{ka\to 0}\left(Q-\frac{1}{B}\right) = \frac{\pi\{\mathrm{Cin}\,\pi-\mathrm{Cin}\,2\pi-\ln(e^{\frac{1}{2}}/4)\}}{\mathrm{Cin}\,2\pi} = 0{\cdot}1249, \tag{3.49}$$

which is small but not zero. On the other hand, the *asymptotic* reciprocal relationship that is under investigation here is seen to hold exactly: the value of Q increases logarithmically with decreasing values of ka while the value of B decreases logarithmically in such a way that their product approaches unity,

$$QB \sim 1. \tag{3.50}$$

Thus, the result of this particular test of the asymptotic reciprocal relationship between bandwidth and Q of radiating systems is clearly affirmative.

Even though the limiting value (3.49) of the difference between these unbounded values of Q and $1/B$ is not zero it is still remarkably small when one considers that it was obtained on the assumption that the current distribution is simply that of a filamentary dipole. This suggests the possibility that the bandwidth and Q of planar antennas, when expressed as they were here in terms of observable stored energies, might have an even closer relationship than the asymptotic reciprocal relationship (3.50) usually attributed to them.

CHAPTER 4

RECTANGULAR AND CIRCULAR SOURCES[9]

THE shape and size of any planar source determines the particular class of functions that can be used to represent physically admissible pattern space factors. Functions of this class were called aperture-limited because of the fact that they are the Fourier transform of functions that vanish outside of the aperture. Each of the two rectangular components (2.13) of the pattern space factor is an aperture-limited function, and although the third component is not it is related to the two that are by (2.15). An essential part of the problem of synthesizing planar sources is to find practical mathematical methods for constructing functions of this class. In the case of all nonrectangular sources, however, the two aperture-limited components cannot be constructed independently of each other because, in general, they are coupled together through the source. The coupling is caused by the form of the edge behaviour that all physical sources were shown in section 2.8 to possess. Whenever the source and its pattern space factor are resolved into components that are nonrectangular each component of the pattern space factor will almost always depend on not just one but both components of the source simultaneously. This coupling appears to be unavoidable for all but rectangular sources, in general, because that is the only shape for which the source can be resolved into rectangular components whose edge exponents $\alpha_{\perp}$ and $\alpha_{\parallel}$ are independent of position.

The effects of coupling will be demonstrated here for the particular case of circular sources. Only in the very special case of circularly symmetric sources will the two components be found to decouple. Although the circular source is the simplest of all nonrectangular sources the effects of coupling are formidable. In view of the complications that they introduce into the synthesis problem the case of circular sources will not be pursued further in this treatise.

Even the relatively simple case of rectangular sources, however, is not entirely free of difficulty. The shape of the visible region of the pattern space factor is always circular, whereas the orthogonality properties of pattern basis functions whose aperture basis functions are orthogonal over the rectangular aperture are themselves restricted to rectangular regions. An orthogonal expansion of the pattern space factor of a rectangular source would thus be incapable of representing individually the radiative and reactive powers needed for a Q-constrained synthesis solution. For that reason the case of arbitrary rectangular sources will not be pursued further either.

But in the special asymptotic cases of rectangular sources whose width approaches infinity (strip sources) or zero (line sources) the two-dimensional pattern space factor becomes essentially one-dimensional, thereby resulting in a precise and relatively simple representation for the radiative and reactive powers individually. Fortunately these special cases represent reasonably good approximations to a number of antennas that are of practical importance. Hence they will be treated in detail in Chapter 5. Because of their relative simplicity the synthesis techniques developed in the remaining chapters of this treatise will be restricted exclusively to these essentially one-dimensional strip and line sources.

4.1. Aperture edge and space factors

From the form of the edge behaviour (2.67) that all physically realizable planar sources must possess it follows that the rectangular components of all rectangular sources of dimensions a and b (Fig. 4.1a) can be represented by

$$E_x(x, y, 0, k) = E_0\{1-(2x/a)^2\}^{\alpha_\perp}\{1-(2y/b)^2\}^{\alpha_\parallel} f_x(2x/a, 2y/b, k), \quad (4.1a)$$

$$E_y(x, y, 0, k) = E_0\{1-(2x/a)^2\}^{\alpha_\parallel}\{1-(2y/b)^2\}^{\alpha_\perp} f_y(2x/a, 2y/b, k), \quad (4.1b)$$

and that the polar components of all circular sources of radius a (Fig. 4.1b) can be represented by

$$E_\rho(\rho, \chi, 0, k) = E_0\{1-(\rho/a)^2\}^{\alpha_\perp} f_\rho(\rho/a, \chi, k), \quad (4.2a)$$

$$E_\chi(\rho, \chi, 0, k) = E_0\{1-(\rho/a)^2\}^{\alpha_\parallel} f_\chi(\rho/a, \chi, k), \quad (4.2b)$$

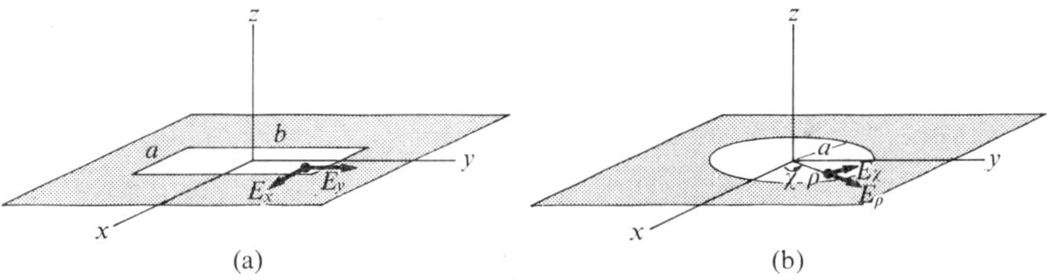

FIG. 4.1. Orthogonal components of planar sources for: (a) rectangular aperture of width a and length b, and (b) circular aperture of radius a.

where the dimensional factor E_0 and the edge exponents $\alpha_\perp$ and $\alpha_\parallel$ are constants. The *edge factor* $\{1-(2x/a)^2\}^{\alpha_\perp}\{1-(2y/b)^2\}^{\alpha_\parallel}$ in (4.1*a*) and $\{1-(2x/a)^2\}^{\alpha_\parallel}\{1-(2y/b)^2\}^{\alpha_\perp}$ in (4.1*b*), and $\{1-(\rho/a)^2\}^{\alpha_\perp}$ in (4.2*a*) and $\{1-(\rho/a)^2\}^{\alpha_\parallel}$ in (4.2*b*), provides a complete description of the edge behaviour. It is independent of the wavenumber k, as required. The dependence upon k and upon the spatial extent of the aperture resides solely in the two dimensionless components $f_x(2x/a, 2y/b, k)$ and $f_y(2x/a, 2y/b, k)$ in (4.1), and

$f_\rho(\rho/a, \chi, k)$ and $f_\chi(\rho/a, \chi, k)$ in (4.2), of the *aperture space factor*. Thus, the aperture distribution for either rectangular or circular apertures always separates into a fixed edge factor and a variable aperture space factor, analogous to the way in which the radiation pattern was shown in section 2.4 to separate into a fixed element factor and a variable pattern space factor. The functional dependence of the radiation pattern upon the aperture distribution is then seen to reside, specifically, in the functional dependence of the pattern space factor upon the aperture space factor.

The two orthogonal components of the aperture space factor for either rectangular or circular sources are dimensionless complex functions whose real and imaginary parts terminate in a pedestal at the edges and are well behaved within the aperture (bounded and continuous, with continuous derivatives) but that appear to be arbitrary otherwise. No special problems occur at the four corners of rectangular apertures because both of the field components must vanish as a corner is approached from any direction.

Since the edge and element factors are both fixed it is concluded that the objective of any source synthesis problem is to find the aperture space factor whose pattern space factor gives a best approximation to some desired pattern function.

4.2. Rectangular aperture-limited functions

A simple relationship exists between the rectangular components of the aperture and pattern space factors of rectangular sources,

$$F_x(k_x, k_y, k) = \tfrac{1}{4}E_0 ab \int_{-1}^{1}\int_{-1}^{1}(1-t_x^2)^{\alpha_\perp}(1-t_y^2)^{\alpha_\parallel} f_x(t_x, t_y, k)\times \times e^{i\{(\frac{1}{2}k_x a)t_x+(\frac{1}{2}k_y b)t_y\}}\,dt_x\,dt_y, \quad (4.3a)$$

$$F_y(k_x, k_y, k) = \tfrac{1}{4}E_0 ab \int_{-1}^{1}\int_{-1}^{1}(1-t_x^2)^{\alpha_\parallel}(1-t_y^2)^{\alpha_\perp} f_y(t_x, t_y, k)\times \times e^{i\{(\frac{1}{2}k_x a)t_x+(\frac{1}{2}k_y b)t_y\}}\,dt_x\,dt_y, \quad (4.3b)$$

obtained by substituting (4.1) into (2.13) and making an obvious change of variables. For an arbitrary aperture space factor this integral represents a special subclass of aperture-limited functions whose behaviour in the neighbourhood of the singularity at infinity is determined by the values of $\alpha_\perp$ and $\alpha_\parallel$. It includes all pattern space factors that are realizable physically for any rectangular source of dimensions a and b with edge exponents $\alpha_\perp$ and $\alpha_\parallel$. Functions of this class are *rectangular aperture-limited functions*. The Fourier transform of such functions is guaranteed to vanish outside the aperture and to behave at the edges according to the values assigned to $\alpha_\perp$ and $\alpha_\parallel$.

4.3. Characteristic functions of rectangular sources

Any two-dimensional square-integrable function can be represented with zero mean-square error over a rectangle by expanding it in a series of products of two complete sets of one-dimensional functions.† The most natural one-dimensional functions for rectangular sources are the characteristic solutions of the integral equation

$$\nu_{\alpha n}(c)\psi_{\alpha n}(c,\eta) = \int_{-1}^{1}(1-t^2)^{\alpha}\psi_{\alpha n}(c,t)e^{icnt}\,dt \tag{4.4}$$

for any $\alpha > -1$, with characteristic numbers $\{\nu_{\alpha n}(c)\}$ indexed by $n = 0, 1, 2, 3, \ldots$. These functions $\{\psi_{\alpha n}(c,\eta)\}$ are the spheroidal functions defined by Stratton.‡ A number of their most important properties are developed in Appendix 3.

Any aperture space factor with rectangular components $f_x(t_x, t_y, k)$ and $f_y(t_x, t_y, k)$ at a fixed value of the wavenumber k can be expanded over the rectangular aperture $-1 \leq t_x \leq 1$, $-1 \leq t_y \leq 1$ in products of spheroidal functions of orders $\alpha_\perp$ and $\alpha_\parallel$,

$$f_x(t_x, t_y, k) = \sum_{n=0}^{\infty}\sum_{p=0}^{\infty}\frac{a_{xnp}}{\nu_{\alpha_\perp n}(c_x)\nu_{\alpha_\parallel p}(c_y)}\psi_{\alpha_\perp n}(c_x, t_x)\psi_{\alpha_\parallel p}(c_y, t_y), \tag{4.5a}$$

$$f_y(t_x, t_y, k) = \sum_{n=0}^{\infty}\sum_{p=0}^{\infty}\frac{a_{ynp}}{\nu_{\alpha_\parallel n}(c_x)\nu_{\alpha_\perp p}(c_y)}\psi_{\alpha_\parallel n}(c_x, t_x)\psi_{\alpha_\perp p}(c_y, t_y), \tag{4.5b}$$

where the parameters c_x and c_y are constants that can be chosen at will. If the values of c_x and c_y are chosen to be

$$c_x = \tfrac{1}{2}ka, \qquad c_y = \tfrac{1}{2}kb, \tag{4.6}$$

then the rectangular components (4.3) of the pattern space factor will be represented exactly by the expansion

$$F_x(k_x, k_y, k) = \tfrac{1}{4}E_0ab\sum_{n=0}^{\infty}\sum_{p=0}^{\infty}a_{xnp}\psi_{\alpha_\perp n}(c_x, k_x/k)\psi_{\alpha\;p}(c_y, k_y/k), \tag{4.7a}$$

$$F_y(k_x, k_y, k) = \tfrac{1}{4}E_0ab\sum_{n=0}^{\infty}\sum_{p=0}^{\infty}a_{ynp}\psi_{\alpha_\parallel n}(c_x, k_x/k)\psi_{\alpha_\perp p}(c_y, k_y/k), \tag{4.7b}$$

† R. Courant and D. Hilbert. *Methods of mathematical physics*, vol. 1, Interscience, New York, 1953; p. 56.

‡ J. A. Stratton. Spheroidal functions. *Proc. natn. Acad. Sci. USA*, **21**, pp. 51–6, January 1935.

after integrating the series (4.5) term by term and using the integral representation (4.4). Thus, the product functions

$$\left.\begin{aligned}\psi_{\alpha_\perp n}(c_x, \eta_x)\psi_{\alpha_\| p}(c_y, \eta_y)\\ \psi_{\alpha_\| n}(c_x, \eta_x)\psi_{\alpha_\perp p}(c_y, \eta_y)\end{aligned}\right\} n,\ p = 0, 1, 2, 3, \ldots, \quad \begin{aligned}(4.8a)\\(4.8b)\end{aligned}$$

can be considered to be the *characteristic functions of rectangular sources* in the sense that the aperture and pattern space factors can be expanded in them simultaneously.

The spheroidal functions form a complete set with respect to all functions that are square-integrable with weight-factor $(1-\eta^2)^\alpha$ on the interval $-1 < \eta < 1$. This completeness property leads to the remarkable conclusion that the very same product functions can be used to expand any pattern function that may be given over the visible region alone, whether aperture-limited or not, since the visible region is bounded by a circle that lies wholly within the rectangle of completeness $-k < k_x < k$, $-k < k_y < k$ of the product functions. On the other hand, the domain of orthogonality of these product functions is the rectangle, not the circle. It is partly this difference between the circular shape of the visible region and the rectangular shape of the domain of orthogonality, and partly a difference between the weighting factors required in the integrals that define the quality factor Q and those that define the orthogonality properties, that destroys the potential usefulness of these characteristic functions for synthesizing Q-constrained rectangular sources. The quality factor Q for all planar sources has been seen to involve integrals of the square of the pattern space factor over the visible and invisible regions separately. If the domain of orthogonality of the characteristic functions had coincided with the shape of the visible region and if, further, the weighting factors in the orthogonality integrals had coincided with those required by the radiative and reactive power integrals then the expression for Q that would be obtained from the pattern series (4.7) would reduce from a ratio of double sums to a ratio of single sums. Such a simplification would lead to optimum Q-constrained solutions for rectangular sources that are similar to those that will be obtained in Chapters 8 and 9 for strip and line sources. But without it the synthesis problem for rectangular sources appears to be formidable indeed.

4.4. Circular aperture-limited functions

To go beyond the case of rectangular sources is still more difficult. The integral representation (2.13) for the two aperture-limited rectangular components of the pattern space factor in terms of rectangular components of the source denotes a relationship between two vector functions that lie in completely independent vector spaces. Such a relationship is valid, in general, only when both vector functions are resolved, as they were there, into rectangular components. But rectangular components of a nonrectangular source are neither parallel nor perpendicular to the edges of the source, from which it follows that the boundary conditions at the aperture edges can no longer be satisfied by each of the rectangular components individually. Each component of the pattern space factor must then be represented by a combination of both components of the source. This complicates the relationship between the aperture and pattern space factors considerably beyond the scalar approximation usually assumed, because it causes the two components of the pattern space factor to become coupled together. The coupling is an inevitable consequence of the vector nature of electromagnetism.

(i) *Arbitrary circular sources*

Changing from rectangular to polar coordinates in the circular aperture shown in Fig. 4.1b,

$$x = \rho \cos \chi, \qquad y = \rho \sin \chi, \tag{4.9}$$

and from rectangular to polar components of the aperture distribution,

$$E_x(x, y, 0, k) = E_\rho(\rho, \chi, 0, k)\cos \chi - E_\chi(\rho, \chi, 0, k)\sin \chi, \tag{4.10a}$$

$$E_y(x, y, 0, k) = E_\rho(\rho, \chi, 0, k)\sin \chi + E_\chi(\rho, \chi, 0, k)\cos \chi, \tag{4.10b}$$

the polar components of the pattern space factor are found to be

$$F_\kappa(\kappa, \phi, k) = \int_0^a \int_{-\pi}^{\pi} \{E_\rho(\rho, \chi, 0, k)\cos(\chi-\phi) - \\ - E_\chi(\rho, \chi, 0, k)\sin(\chi-\phi)\}e^{i\kappa\rho\cos(\chi-\phi)}\rho\, d\rho\, d\chi, \tag{4.11a}$$

$$F_\phi(\kappa, \phi, k) = \int_0^a \int_{-\pi}^{\pi} \{E_\rho(\rho, \chi, 0, k)\sin(\chi-\phi) + \\ + E_x(\rho, \chi, 0, k)\cos(\chi-\phi)\}e^{i\kappa\rho\cos(\chi-\phi)}\rho\, d\rho\, d\chi, \tag{4.11b}$$

from (2.13) and (2.38). The aperture edge condition for physical realizability can now be imposed. It requires that each of the two polar components of the aperture distribution contain an edge factor as in (4.2). Upon recognizing that the angular dependence of the aperture space factor must always be periodic with period 2π, and so can always be expanded in a Fourier series,

$$f_\rho(\rho/a, \chi, k) = \sum_{m=-\infty}^{\infty} g_{\rho m}(\rho/a, k)e^{im\chi}, \tag{4.12a}$$

$$f_\chi(\rho/a, \chi, k) = \sum_{m=-\infty}^{\infty} g_{\chi m}(\rho/a, k)e^{im\chi}, \tag{4.12b}$$

the integrals in (4.11) can be evaluated term by term, the integration on χ in each term resulting in Bessel functions:

$$F_{\kappa}(\kappa, \phi, k) = 2\pi E_0 a^2 \sum_{m=-\infty}^{\infty} i^m e^{im\phi} \left\{ -i \frac{\partial G_{\rho\alpha_\perp m}(\kappa a, k)}{\partial(\kappa a)} - m \frac{G_{\chi\alpha_\| m}(\kappa a, k)}{\kappa a} \right\}, \quad (4.13a)$$

$$F_{\phi}(\kappa, \phi, k) = 2\pi E_0 a^2 \sum_{m=-\infty}^{\infty} i^m e^{im\phi} \left\{ m \frac{G_{\rho\alpha_\perp m}(\kappa a, k)}{\kappa a} - i \frac{\partial G_{\chi\alpha_\| m}(\kappa a, k)}{\partial(\kappa a)} \right\}, \quad (4.13b)$$

where

$$G_{\rho\alpha_\perp m}(\kappa a, k) = \int_0^1 (1-t^2)^{\alpha_\perp} g_{\rho m}(t, k) J_m(\kappa a t)\, dt, \quad (4.14a)$$

$$G_{\chi\alpha_\| m}(\kappa a, k) = \int_0^1 (1-t^2)^{\alpha_\|} g_{\chi m}(t, k) J_m(\kappa a t)\, dt. \quad (4.14b)$$

These formulas (4.13) are exact and completely general expressions for any physically realizable pattern space factor of a circular source. They are expressed in terms of two independent sets of one-dimensional *circular aperture-limited functions* $G_{\rho\alpha_\perp m}(\kappa a, k)$ and $G_{\chi\alpha_\| m}(\kappa a, k)$ of orders $\alpha_\perp$ and $\alpha_\|$, respectively, one for each of the two polar components of the aperture space factor.

The relative complexity of circular sources is now fully evident: each component of the pattern space factor depends upon a combination of both components of the aperture space factor, in general, whereas for rectangular sources it depends upon only one. This is the way that the two are coupled together.

It is interesting to compare these exact results for vector sources with the corresponding results for scalar sources; e.g. for acoustic sources. Any scalar source $E(\rho, \chi, 0, k)$ can again be represented by the product of a dimensional constant E_0 by an edge factor $\{1-(\rho/a)^2\}^\alpha$ and an aperture space factor $f(\rho/a, \chi, k)$,

$$f(\rho/a, \chi, k) = \sum_{m=-\infty}^{\infty} g_m(\rho/a, k) e^{im\chi}, \quad (4.15)$$

where the value of α can be chosen in such a way as to obtain any prescribed edge taper. The corresponding scalar pattern space factor $F(\kappa, \phi, k)$ will then be

$$F(\kappa, \phi, k) = \int_0^a \int_{-\pi}^{\pi} E(\rho, \chi, 0, k) e^{i\kappa\rho\cos(\chi-\phi)} \rho\, d\rho\, d\chi$$

$$= 2\pi E_0 a^2 \sum_{m=-\infty}^{\infty} i^m e^{im\phi} \int_0 (1-t^2)^\alpha g_m(t, k) J_{m)}\kappa a t(t\, dt. \quad (4.16)$$

This is much simpler than either of the two coupled scalar components (4.13) for vector sources.

(ii) *Circularly symmetric sources*

There appears to be but one type of circular source for which the two components (4.13) of the vector pattern space factor are decoupled. This is the class of all circularly symmetric sources (i.e. those for which $g_{\rho m}(\rho/a, k)$ and $g_{\chi m}(\rho/a, k)$ are zero for all $m \neq 0$). The two polar components (4.13) of the pattern space factor are then also circularly symmetric, and hence decouple, reducing to

$$F_{\kappa}(\kappa, \phi, k) = i2\pi E_0 a^2 \int_0^1 (1-t^2)^{\alpha_\perp} g_{\rho 0}(t, k) J_1(\kappa a t) t \, dt, \tag{4.17a}$$

$$F_{\phi}(\kappa, \phi, k) = i2\pi E_0 a^2 \int_0^1 (1-t^2)^{\alpha_\parallel} g_{\chi 0}(t, k) J_1(\kappa a t) t \, dt. \tag{4.17b}$$

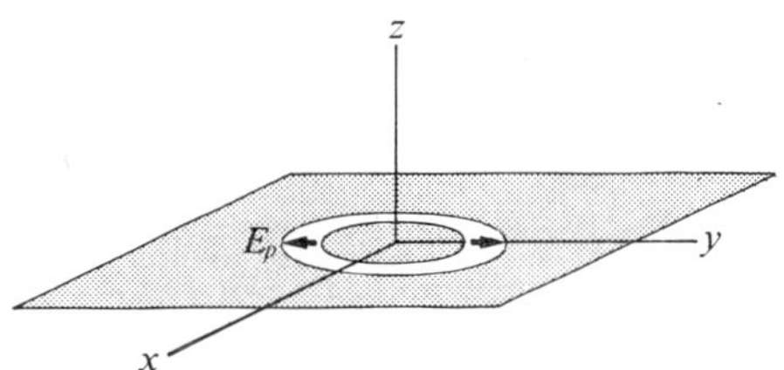

FIG. 4.2. Annular slot with purely radial electric field.

In this case the elevation component $F_\kappa(\kappa, \phi, k)$ depends upon only $g_{\rho 0}(t, k)$ and the azimuthal component $F_\phi(\kappa, \phi, k)$ depends upon only $g_{\chi 0}(t, k)$.

This relationship between the aperture and pattern space factors of vector sources with circular symmetry is quite different from the corresponding one ($m = 0$) that would be obtained from (4.16) for scalar sources. The difference arises from the Bessel function $J_1(\kappa a t)$ that appears in place of $J_0(\kappa a t)$. One consequence of this difference is that the broadside value of the radiation pattern of any circularly symmetric vector source will always be zero instead of a maximum. For this reason the special case of circular symmetry for true vector sources is believed to be of limited practical interest.

One important practical example of a planar antenna whose source is circularly symmetric is the annular slot with purely radial electric field (Fig. 4.2). Such an antenna acts as though it were a magnetic loop, producing zero field on its axis. Its pattern space factor has only a radial component $F_\kappa(\kappa, \phi, k)$, which means that its observable stored energy must be purely electric since its observable magnetic energy (3.40*a*) is always exactly zero.

4.5. Characteristic functions of circular sources

Another consequence of the coupling that exists between components of the pattern space factor is that, in general, there are no functions that are characteristic of circular sources in the sense that the aperture and pattern space factors can be expanded in them simultaneously. The only exception appears to be in the special case of circular symmetry. For scalar sources, on the other hand, there is a set of characteristic functions that can represent completely arbitrary aperture and pattern space factors for arbitrary values of the edge exponent $\alpha > -1$.

In the case of circularly symmetric vector sources the Fourier-Bessel transform in (4.17) for any given value of the wavenumber k is a one-dimensional function of the form

$$G_{\alpha\mu}(c\eta) = \int_0^1 (1-t^2)^{\alpha} g(t) J_{\mu}(c\eta t) t \, dt \tag{4.18}$$

for $\mu = 1$, where $c = ka$ and $\eta = \kappa/k$. By investigating the properties of this integral as a function of the arbitrary parameter $\mu > -1$ all conclusions drawn for the pattern function (4.17) of circularly symmetric vector sources can be extended to include the individual terms for $\mu = m = 0, 1, 2, 3, \ldots$ in the expansion (4.16) for completely arbitrary scalar sources (m can be restricted to the nonnegative integers by proper pairing of terms).

The characteristic functions associated with the transform (4.18) are the characteristic functions of the integral equation

$$\nu_{\alpha\mu n}(c) R_{\alpha\mu n}(c, \eta) = \int_0^1 (1-t^2)^{\alpha} R_{\alpha\mu n}(c, t) J_{\mu}(c\eta t) t \, dt \tag{4.19}$$

for $\alpha > -1$ and $\mu > -1$. The characteristic values $\{\nu_{\alpha\mu n}(c)\}$ are indexed by $n = 0, 1, 2, 3, \ldots$. This integral equation is related to a generalization to arbitrary real values $\alpha > -1$ of a simpler integral equation (for $\alpha = 0$) investigated recently by Slepian.† It is also related to a generalization to arbitrary real values $\mu > -1$ of the simpler integral equation (4.4) (for $\mu = \pm\frac{1}{2}$) investigated by Stratton. The relationship of the characteristic functions $\{R_{\alpha\mu n}(c, \eta)\}$ to, and some of the pertinent properties of, the generalized spheroidal functions are developed in Appendix 4.

The aperture function $g(t)$ can be expanded in the orthogonal set of characteristic functions $\{R_{\alpha\mu n}(c, t)\}$ over the interval (0, 1),

$$g(t) = \sum_{n=0}^{\infty} \frac{a_n}{\nu_{\alpha\mu n}(c)} R_{\alpha\mu n}(c, t) \tag{4.20}$$

† D. Slepian. Prolate spheroidal wave functions, Fourier analysis, and uncertainty—IV, *Bell System Tech. J.*, **43**, pp. 3009–58, November 1964.

for any $\alpha > -1$ and $\mu > -1$. Hence the function $G_{\alpha\mu}(c\eta)$ will be represented everywhere in η by an expansion in the same functions,

$$G_{\alpha\mu}(c\eta) = \sum_{n=0}^{\infty} a_n R_{\alpha\mu n}(c, \eta), \tag{4.21}$$

by integrating the series (4.20) term by term and using (4.19). For circularly symmetric vector sources, then, the aperture space factor (4.12) and the pattern space vector (4.17) can be expanded simultaneously in a single set of functions,

$$f_\rho(\rho/a, \chi, k) = \sum_{n=0}^{\infty} \frac{a_{\rho n}}{\nu_{\alpha_\perp 1n}(ka)} R_{\alpha_\perp 1n}(ka, \rho/a), \tag{4.22a}$$

$$f_\chi(\rho/a, \chi, k) = \sum_{n=0}^{\infty} \frac{a_{\chi n}}{\nu_{\alpha_\parallel 1n}(ka)} R_{\alpha_\parallel 1n}(ka, \rho/a), \tag{4.22b}$$

and

$$F_\kappa(\kappa, \phi, k) = i2\pi E_0 a^2 \sum_{n=0}^{\infty} a_{\rho n} R_{\alpha_\perp 1n}(ka, \kappa/k), \tag{4.23a}$$

$$F_\phi(\kappa, \phi, k) = i2\pi E_0 a^2 \sum_{n=0}^{\infty} a_{\chi n} R_{\alpha_\parallel 1n}(ka, \kappa/k). \tag{4.23b}$$

For completely arbitrary scalar sources with edge exponent α the aperture space factor (4.15) and its pattern space factor (4.16), obtained by first pairing the terms for $\pm m$ and then expanding the sum and the difference of $g_m(t, k)$ and $g_{-m}(t, k)$ in the series

$$g_m(t, k) + g_{-m}(t, k) = \frac{2}{\epsilon_m} \sum_{n=0}^{\infty} \frac{a_{mn}}{\nu_{\alpha mn}(c)} R_{\alpha mn}(c, t) \tag{4.24a}$$

$$g_m(t, k) - g_{-m}(t, k) = \sum_{n=0}^{\infty} \frac{b_{mn}}{i\nu_{\alpha mn}(c)} R_{\alpha mn}(c, t) \tag{4.24b}$$

for $m = 0, 1, 2, 3, \ldots$, where ϵ_m denotes 1 for $m = 0$ and 2 for $m \neq 0$ (Neumann's number), can also be expanded simultaneously in a single set of functions,

$$f(\rho/a, \chi, k) = \sum_{m=0}^{\infty} \sum_{n=0}^{\infty} \frac{a_{mn} \cos m\chi + b_{mn} \sin m\chi}{\nu_{\alpha mn}(ka)} R_{\alpha mn}(ka, \rho/a), \tag{4.25}$$

$$F(\kappa, \phi, k) = 2\pi E_0 a^2 \sum_{m=0}^{\infty} \sum_{n=0}^{\infty} i^m (a_{mn} \cos m\phi + b_{mn} \sin m\phi) R_{\alpha mn}(ka, \kappa/k). \tag{4.26}$$

Thus, the *characteristic functions of circularly symmetric vector sources* (the only circular vector sources for which characteristic functions are known to exist) are seen to be

$$R_{\alpha 1n}(ka, \eta), \qquad n = 0, 1, 2, 3, \ldots \tag{4.27}$$

On the other hand, the *characteristic functions of circular scalar sources* are seen to be

$$\left.\begin{matrix}\cos\\ \sin\end{matrix}\right\} m\phi R_{\alpha mn}(ka, \eta), \qquad m, n = 0, 1, 2, 3, \ldots \tag{4.28}$$

The latter reduce to Slepian's circular eigenfunctions† in the special case where the edge exponent has the particular value $\alpha = 0$; i.e. where the scalar source terminates abruptly in a pedestal at the circular edge.

† D. Slepian, *loc. cit.*, eqn (17)

CHAPTER 5
STRIP AND LINE SOURCES[11]

No optimum planar source of any kind has yet been synthesized in two dimensions and none is foreseen for the immediate future. Not for lack of interest, however, but because there are still a number of rather severe mathematical difficulties to be overcome. In the case of all nonrectangular sources there are the problems that arise from the coupling that almost always occurs between components of the pattern space factor. The special case of circularly symmetric sources that was treated in Chapter 4 is the only type of nonrectangular planar source known whose pattern components are not coupled together. And even that special case can be synthesized in only one dimension (radial) because it is not free to vary in the other (angular). In the case of rectangular sources there is no coupling between components but there are still the problems that arise from the circular shape of the visible region of the pattern space factor. The only rectangular sources known to be free of the difficulties caused by coupling and by the circular shape of the visible region are strip and line sources. Like circularly symmetric sources, however, strip and line sources can be synthesized in only one dimension (length) because they are not free to vary in the other (width).

Strip and line sources appear to provide the principal class of physically interesting synthesis problems that are capable of being solved by existing mathematical methods. For that reason their analytical properties will be developed here in some detail. These properties will provide the physical basis for the synthesis techniques developed in the concluding four chapters.

5.1. Definition of strip and line sources

By the term 'strip source' will be meant specifically a rectangular source whose width approaches infinity and whose aperture space factor is uniform across the width. It will be called an *E-plane strip source* if the tangential electric field in the aperture is directed entirely along the length of the aperture, and an *H-plane strip source* if the tangential electric field is directed entirely along the width. Any other strip source can be obtained by superposing an appropriate combination of E- and H-plane strip sources.

By the term 'line source' will be meant specifically a rectangular source whose aperture space factor is again uniform across the width but whose width approaches zero rather than infinity. The only nonvanishing tangential component of electric field that can exist in an aperture of vanishing width is the one across the width, hence all line sources are exclusively *H-plane line sources*. The complement to the filamentary dipole antenna treated in Chapter 3 is an example of a practical line source.

All of the strip and line sources considered in this treatise will be assumed to have but a single component $E_x(x, y, 0, k)$. Denoting the dimensions of the rectangular source by a and b as in Fig. 5.1a, H-plane strip and line sources are obtained by letting the width a approach infinity (Fig. 5.1b) and zero

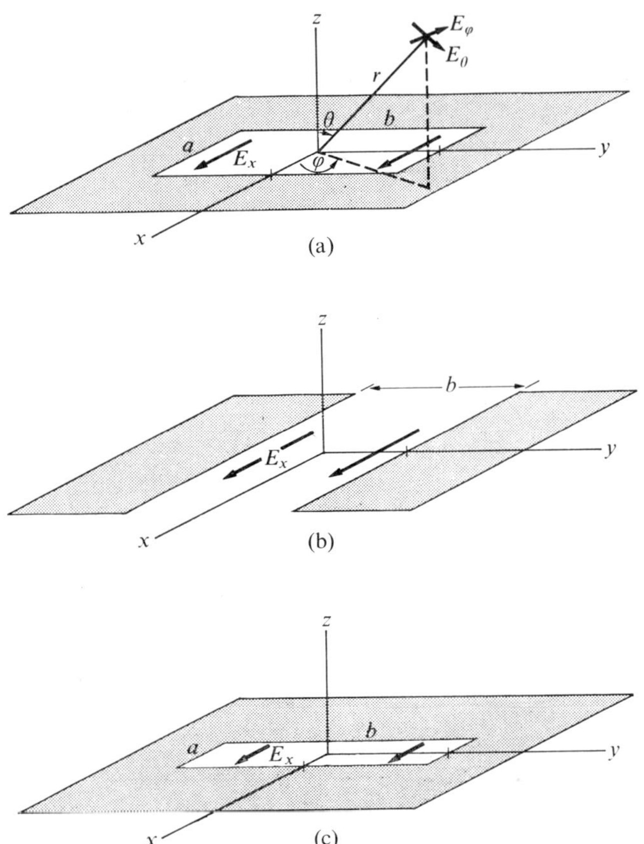

FIG. 5.1. (a) Rectangular source $E_x(x, y, 0, k)$ of arbitrary dimensions a and b; (b) limiting case of H-plane strip source ($a \to \infty$); and (c) limiting case of H-plane line source ($a \to 0$).

(Fig. 5.1c), respectively. Similarly, E-plane strip sources are obtained by letting the width b approach infinity.

Strip and line sources are obviously not realizable physically because there are practical limits on how large and how small the width of physical sources can be made. But it is important to recognize that they do represent useful approximations to certain physical sources that are of considerable practical importance. Strip sources provide a reasonable approximation to all rectangular sources that are highly directive in one principal plane; e.g. they can be used to represent the aperture distribution for the narrow fan-shaped beam required for search radars. Line sources, on the other hand, provide a reasonable approximation to a wide variety of sources whose radiation pattern is nearly a surface of revolution about the axis of the source; e.g., narrow slot and dipole antennas, long wire antennas, broadcast towers, etc.

5.2. Aperture and pattern space factors

The aperture space factor of a strip or line source, obtained from the aperture space factor $f_x(2x/a, 2y/b, k)$ of the rectangular source (4.1) at a fixed value of the wavenumber k, is uniform over the width, varying only in the x-direction for E-plane sources and in the y-direction for H-plane sources. Hence it can be represented simply as a one-dimensional function,

$$f_x(2x/a, 2y/b, k) = \begin{cases} f_{xx}(2x/a), & E\text{-plane} \\ f_{xy}(2y/b), & H\text{-plane}\,. \end{cases} \tag{5.1}$$

The pattern space factor (4.3) then becomes a separable function of k_x and k_y,

$$F_x(k_x, k_y, k) = \begin{cases} W_{\alpha_{\parallel}}(\tfrac{1}{2}k_y b)F_{xx}(\tfrac{1}{2}k_x a), & E\text{-plane} \\ W_{\alpha_{\perp}}(\tfrac{1}{2}k_x a)F_{xy}(\tfrac{1}{2}k_y b), & H\text{-plane} \end{cases} \tag{5.2}$$

in which the pattern in the principal plane of the width is found, after carrying out the integration, to be

$$W_{\alpha_{\perp}}(c_x n_x) = \tfrac{1}{4}E_0 ab 2^{\alpha_{\perp}+\frac{1}{2}}\Gamma(\alpha_{\perp}+1)\Gamma(\tfrac{1}{2})\frac{J_{\alpha_{\perp}+\frac{1}{2}}(c_x n_x)}{(c_x n_x)^{\alpha_{\perp}+\frac{1}{2}}}, \tag{5.3a}$$

$$W_{\alpha_{\parallel}}(c_y n_y) = \tfrac{1}{4}E_0 ab 2^{\alpha_{\parallel}+\frac{1}{2}}\Gamma(\alpha_{\parallel}+1)\Gamma(\tfrac{1}{2})\frac{J_{\alpha_{\parallel}+\frac{1}{2}}(c_y n_y)}{(c_y n_y)^{\alpha_{\parallel}+\frac{1}{2}}}, \tag{5.3b}$$

and in which the pattern in the principal plane of the length is either

$$F_{xx}(c_x n_x) = \int_{-1}^{1}(1-t_x^2)^{\alpha_{\perp}} f_{xx}(t_x)e^{ic_x n_x t_x}\,dt_x, \tag{5.4a}$$

or

$$F_{xy}(c_y n_y) = \int_{-1}^{1}(1-t_y^2)^{\alpha_{\parallel}} f_{xy}(t_y)e^{ic_y n_y t_y}\,dt_y. \tag{5.4b}$$

The components k_x and k_y of the vector wavenumber $\mathbf{k}$ have been expressed here in terms of the corresponding components of the unit normal $\mathbf{n}$ (the direction cosines n_x and n_y),

$$n_x = \sin\theta\cos\phi, \qquad n_y = \sin\theta\sin\phi, \tag{5.5}$$

and in terms of the constants

$$c_x = \tfrac{1}{2}ka, \qquad c_y = \tfrac{1}{2}kb, \tag{5.6}$$

that describe the electrical dimensions of the source.

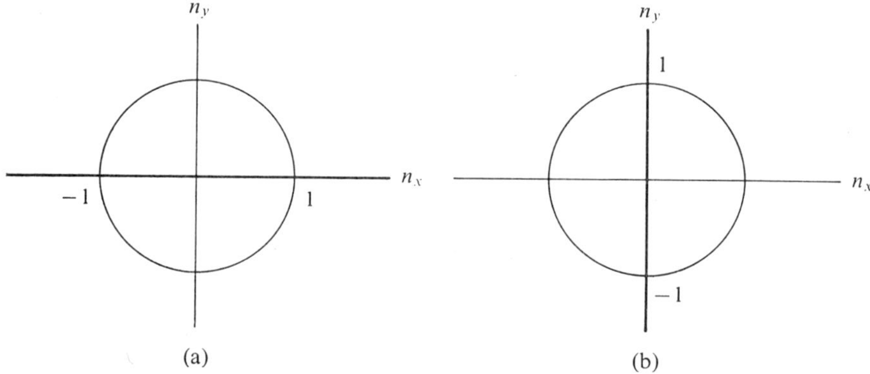

FIG. 5.2. Concentration of the pattern space factor about: (a) the n_x axis for E-plane strip sources; and (b) the n_y axis for H-plane strip sources.

The asymptotic form of the pattern space factor for strip and for line sources will now be found from the asymptotic form of the pattern function (5.3) in the principal plane of the width.

As the width approaches infinity the function (5.3) approaches

$$W_\alpha(cn) \sim \begin{cases} \tfrac{1}{4}E_0ab2^{\alpha+1}\Gamma(\alpha+1)\dfrac{\sin(cn-\frac{1}{2}\alpha\pi)}{(cn)^{\alpha+1}}, & n \neq 0 \\ \tfrac{1}{4}E_0ab\Gamma(\alpha+1)\Gamma(\tfrac{1}{2})/\Gamma(\alpha+\tfrac{3}{2}), & n = 0, \end{cases} \tag{5.7}$$

where cn denotes $c_x n_x$ for $\alpha = \alpha_\perp$ and $c_y n_y$ for $\alpha = \alpha_\parallel$. Thus, for strip sources the two-dimensional pattern space factor $F_\alpha(kn_x, kn_y, k)$ becomes much like a delta function of n_x or n_y for all values of $\alpha > -1$: it concentrates entirely about the n_x axis in the case of E-plane strip sources, as indicated in Fig. 5.2a, and entirely about the n_y axis in the case of H-plane strip sources, as indicated in Fig. 5.2b. The visible region then collapses from the two-dimensional interior of the unit circle in the direction-cosine plane to just the one-dimensional interior of the interval $(-1, 1)$ along the n_x or n_y axis, while the invisible region collapses from the rest of the direction-cosine plane to just the rest of the n_x or n_y axis. The pattern space factor then approaches a truly one-dimensional function like that assumed in the one-dimensional analysis of Woodward and Lawson.†

As the width approaches zero the limiting form of the function (5.3a) is quite different. Instead of approaching a delta function of n_x it spreads out uniformly along the n_x axis, approaching the constant value

$$W_{\alpha_\perp}(c_x n_x) \sim \tfrac{1}{4}E_0 ab\Gamma(\alpha_\perp+1)\Gamma(\tfrac{1}{2})/\Gamma(\alpha_\perp+\tfrac{3}{2}). \tag{5.8}$$

Thus, the one-dimensional H-plane pattern function (5.4*b*) has a distinctly different physical interpretation for line sources than for strip sources. In the case of strip sources there was a sharp separation between the visible and invisible regions, the visible region collapsing to just the interval $-1 < n_y < 1$ along the n_y axis and the invisible region collapsing to just the rest of the line $-\infty < n_y < \infty$. This is not true for line sources because the 'visible' region defined by $-1 < n_y < 1$ now includes the entire circle of the actual visible region and an infinitely long strip of the invisible region as well. This essential difference between the physics of strip and of line sources is one that appears to have gone largely unrecognized. It will be seen in section 5.4 to result in an essential difference in the functional form of the quality factor Q of strip and of line sources.

From (5.4*a*) and (5.4*b*) it is evident that the only difference in the functional form of the pattern space factor of E- and H-plane sources lies in the value assigned to the edge exponent $\alpha_\perp$ and $\alpha_\parallel$, respectively. It will be convenient, therefore, to describe them both by a single function $F_\alpha(c\eta)$,

$$F_\alpha(c\eta) = \int_{-1}^{1}(1-t^2)^\alpha f(t)e^{ic\eta t}\,dt, \tag{5.9}$$

where η denotes the direction cosine and c denotes $kl/2$ for a source of length l (a for E-plane sources, b for H-plane sources). Functions of the class (5.9), in which $f(t)$ is any function that is square-integrable with weight factor $(1-t^2)^\alpha$ on $(-1, 1)$ and that terminates at the edges $t = \pm 1$ in a pedestal, will be called simply *aperture-limited functions of order* α.

The value of the edge exponents $\alpha_\perp$ and $\alpha_\parallel$ at the ends of E- and H-plane sources, respectively, will be assumed to be arbitrary, even though the only value that is exactly realizable physically is $\alpha_\perp = 1$ and $\alpha_\parallel = 2$, from (2.68). Along the infinitely distant edges of E- and H-plane strip sources the value of $\alpha_\parallel$ and $\alpha_\perp$, respectively, that appears in (5.3) can also be assumed to be arbitrary, as all of the succeeding calculations will be seen to be unaffected by it. But in the case of the vanishingly near edges of H-plane line sources it will be seen that the calculation of Q does depend upon the particular value chosen for $\alpha_\perp$. For that case it will be convenient to assign the value

† P. M. Woodward and J. D. Lawson, *loc. cit.*

$\alpha_\perp = 0$ (a pedestal) that is usually implied in most line source calculations, as in the case, for example, of the sinusoidal distribution (3.7) usually assumed for narrow dipole antennas near half-wave resonance.

5.3. Characteristic functions

A convenient analytical method for constructing an arbitrary pattern space factor $F_\alpha(c\eta)$ is to represent it by an expansion in some set of basis functions that is complete with respect to all aperture-limited functions of order α. Basis functions of this type can be obtained by taking the transform (5.9) of any set of aperture basis functions that terminate in a pedestal at the edges $t = \pm 1$ and that are complete on $(-1, 1)$ with respect to all functions that are square-integrable with weight factor $(1-t^2)^\alpha$.

Several different complete sets of aperture basis functions and their transforms have been proposed by various authors. The simplest ones for $\alpha = 0$ or 1 are the sine and cosine functions whose transformed functions are the sampling functions that will be treated in detail in section 6.3. Another is the set of Chebyshev polynomials of the first kind ($\alpha = -\frac{1}{2}$) proposed by Tartakovskii,† for which the transformed functions are Bessel functions of integral order. Still another is the set of Legendre polynomials ($\alpha = 0$) proposed by Minkovich,‡ for which the transformed functions are spherical Bessel functions. A more general set of functions that is valid for any $\alpha > -1$ and that includes both the Chebyshev and the Legendre polynomials as special cases is the Gegenbauer polynomials $\{T_n^\alpha(t)\}$ of order α, whose transformed functions are known§ to be

$$\int_{-1}^{1} (1-t^2)^\alpha T_n^\alpha(t) e^{ic\eta t}\, dt = \frac{(2\pi)^{\frac{1}{2}} i^n \Gamma(n+2\alpha+1)}{n!} \frac{J_{n+\alpha+\frac{1}{2}}(c\eta)}{(c\eta)^{\alpha+\frac{1}{2}}}. \tag{5.10}$$

There is one particular set of aperture basis functions and their transforms, however, that is uniquely suited to the synthesis of strip and line sources. This is the set of spheroidal functions $\{\psi_{\alpha n}(c, \eta)\}$ of arbitrary order $\alpha > -1$ that was introduced briefly into the theory of rectangular sources in section 4.3. When the parameter c approaches zero the spheroidal functions become proportional to the Gegenbauer polynomials of order α on the interval $(-1, 1)$ and their transform becomes proportional to (5.10). But for any positive value of c the spheroidal functions have the distinctive property of being proportional to their own transform. They arise as the characteristic

† L. B. Tartakovskii. The synthesis of a linear radiator and its analogy in the wideband matching problem. *Radio Engng electron. Phys.*, **3**, pp. 73–89, December 1958.

‡ B. M. Minkovich. Concerning one type of partial radiation pattern. *Radio Engng electron. Phys.*, **7**, pp. 666–8, April 1962.

§ P. M. Morse and H. Feshbach. *Methods of theoretical physics*, McGraw-Hill, New York 1953; eqn (5.3.67).

functions of the transformation (5.9) for strip and line sources; they are the characteristic solutions of the spheroidal integral equation,

$$\nu_{\alpha n}(c)\psi_{\alpha n}(c,\eta) = \int_{-1}^{1} (1-t^2)^\alpha \psi_{\alpha n}(c,t)e^{ic\eta t}\,dt. \tag{5.11}$$

Thus, the spheroidal functions $\{\psi_{\alpha n}(c,\eta)\}$ of order α are the *characteristic functions of strip and line sources* in the same sense that the products (4.8) of pairs of spheroidal functions were seen to be the characteristic functions of rectangular sources. But they are also characteristic in quite another, and far more important, sense: they provide pattern basis functions whose contributions (exactly or to a good approximation) to the radiative and reactive powers, separately, are linearly independent. This is a direct consequence of the remarkable fact that the spheroidal functions are *doubly orthogonal*; i.e. they are orthogonal over both the interval $(-1, 1)$ and $(-\infty, \infty)$ simultaneously, as shown in Appendix 3. The two orthonormality integrals are as follows:

$$\int_{-1}^{1} \psi_{\alpha n}(c,\eta)\psi_{\alpha p}(c,\eta)(1-\eta^2)^\alpha\,d\eta = \Lambda_{\alpha n}(c)\,\delta_{np}, \tag{5.12a}$$

$$\int_{-\infty}^{\infty} \psi_{\alpha n}(c,\eta)\psi_{\alpha p}(c,\eta)\,|1-\eta^2|^\alpha\,d\eta = \gamma_{\alpha n}(c)\Lambda_{\alpha n}(c)\,\delta_{np}, \tag{5.12b}$$

where $\{\Lambda_{\alpha n}(c)\}$ denotes an arbitrarily chosen (positive) set of normalization constants on the interval $(-1, 1)$ and where $\{\gamma_{\alpha n}(c)\}$ denotes a third kind of characteristic numbers (also positive) that determine the ratio of the two different sets of normalization constants. In section 5.4 it will be shown that for the case of strip sources the radiative power arises from the interval $(-1, 1)$ and the reactive power from the difference between the two intervals $(-\infty, \infty)$ and $(-1, 1)$, while for the case of line sources the radiative power again arises from the interval $(-1, 1)$ but the reactive power arises instead from the full interval $(-\infty, \infty)$. For either strip or line sources, then, the spheroidal functions are orthogonal over both the region of radiative power and the region of reactive power separately. This results in linear independence of the individual contributions to each of the two powers separately.

In addition to their being doubly orthogonal the spheroidal functions are also shown in Appendix 3 to have the important property of being *doubly complete*: they are complete on $(-1, 1)$ with respect to all functions that are square-integrable with weight factor $(1-\eta^2)^\alpha$, and they are also complete on $(-\infty, \infty)$ with respect to all aperture-limited functions. Hence they are capable of representing exactly any non-aperture-limited pattern function that may be specified over the visible region $(-1, 1)$, in addition to representing exactly its aperture-limited approximation $F_\alpha(c\eta)$ everywhere on the real axis $(-\infty, \infty)$. At the same time they are also capable of representing exactly the aperture space factor $f(t)$ required to produce that aperture-limited approximation. Thus, this one set of functions, alone, can be used to represent all three of the functions that enter into strip and line source synthesis problems.

The double orthogonality and double completeness properties of the spheroidal functions will be seen in Chapters 8 and 9 to play a key role in the synthesis of strip and line sources when optimized either for maximum directivity or for minimum radiated power in the error pattern. These properties were first discovered by Slepian and Pollak† for the special case of $\alpha = 0$ and later generalized by the author[5] to include all spheroidal functions of arbitrary real order $\alpha > -1$.

Another important property of the spheroidal functions of order zero that was first observed by Slepian and Pollak is that, for a fixed value of c, the tabulated values of the reciprocal of $\{\gamma_{\alpha n}(c)\}$ for $\alpha = 0$ ($\{\lambda_n\}$ in their notation) fall off rapidly to zero with increasing n once the value of n has exceeded $2c/\pi$. From the tabulated values of the reciprocal of $\{\gamma_{\alpha n}(c)\}$ shown in Table A3.2 of Appendix 3 for various values of α it appears that this property is true quite generally for all orders $\alpha > -1$. Its importance to the theory of source synthesis lies in the fact that the reciprocal of $\{\gamma_{\alpha n}(c)\}$ is a measure of the relative concentration of power within the visible region $(-1, 1)$ of each spheroidal function (see section A3.5). When the value of n exceeds $2c/\pi$ (or $2l/\lambda$) the power in successive functions is pushed more and more into the invisible region. In any Q-constrained synthesis solution, then, the coefficients of these higher terms in the pattern series must fall rapidly to zero. Thus, the series expansion of any Q-constrained pattern space factor will exhibit a peculiar convergence property, in that the value of the series will always be determined largely by the first $(2l/\lambda)+1$ terms (one more than twice the number of wavelengths in the aperture) beyond which the series will converge quickly for all practical values of Q. It is of interest to note that this is essentially the same as the number of $\{\sin(u-n\pi)\}/(u-n\pi)$ terms that Woodward used (one more than twice the largest whole number of wavelengths in the aperture) in a sampling synthesis method that will be described later in section 6.1.

Several sets of spheroidal functions $\{\psi_{\alpha n}(c, \eta)\}$ for various values of α are illustrated in Fig. A3.1 of Appendix 3 for $c = 6$. Spheroidal functions are even functions of η for even n and odd functions of η for odd n, hence they are shown for positive η only. Each function has three distinctly different kinds of characteristic numbers associated with it, numerical examples of which are shown in Figs. A3.2–A3.4 and in Tables A3.1 and A3.2.

† D. Slepian and H. O. Pollak. Prolate spheroidal wave functions, Fourier analysis, and uncertainty—I. *Bell System Tech. J.*, **40**, pp. 43–64, January 1961.

Two important special cases are the even and odd (as a function of arc cos η) periodic Mathieu functions. The even Mathieu functions $\{Se_n(c, \eta)\}$ are precisely the spheroidal functions of order $\alpha = -\frac{1}{2}$, and the odd Mathieu functions $\{So_{n+1}(c, \eta)\}$ are $(1-\eta^2)^{\frac{1}{2}}$ times the spheroidal functions of order $\alpha = +\frac{1}{2}$. Mathieu functions had been proposed earlier by Pistolkors† and by Leonard‡ as basis functions for synthesizing strip sources but neither Pistolkors nor Leonard appears to have been aware of the double orthogonality property. And nothing was yet known about the formulation of Q as a constraint upon synthesis solutions. The formulation of Q as a constraining parameter for both strip and line sources will be developed in the next section.

5.4. Quality factor Q

In section 3.4 it was shown that for the meaning of Q to be consistent with bandwidth it must be expressed in terms of the dissipative power and the total observable reactive power that would be seen at the input port of a complete antenna system. But a planar source does not constitute a complete antenna system, in general, because it does not include the cavity required to excite it. Hence the radiative power (3.5) and the observable reactive powers (3.38) of planar sources are but an incomplete representation for the dissipative and observable reactive powers that would be seen at the input port of a complete planar antenna. The part that is missing is the dissipative power and the observable reactive powers that would be contributed from within the exciting cavity.

Without specific knowledge of the dissipative power and the observable reactive powers contributed by the cavity itself the constraint imposed by Q upon the synthesis of radiating sources is only incompletely defined. It is important to recognize, and indeed can hardly be emphasized strongly enough, that all constrained source synthesis problems, including the synthesis of strip and line sources that will be treated in Chapters 8 and 9, are only incompletely defined problems for this very reason. In order to complete the definition of the source synthesis problem, then, some assumption will have to be made regarding the contributions made by the cavity. Hence, it will be assumed throughout the rest of this treatise that *all observable effects of the cavity as seen at its input port can be replaced by a lumped inductor or capacitor that resonates with the source.* This is equivalent, in effect, to assuming that the total observable stored energy attributable to the cavity itself is purely magnetic or electric, respectively, and that the cavity is entirely lossless. All of the properties obtained henceforth for strip and line source

† A. A. Pistolkors. Use of Mathieu functions for computing field distribution in an antenna to obtain a given directional diagram. *Doklady AN SSSR* (*English translation*), **89**, pp. 849–52, 1953.

‡ D. J. Leonard. The synthesis of the aperture field from an arbitrary radiation pattern. *McGill Symposium on microwave optics* (1953), *Part II*, pp. 337–43; 1959.

antennas will be assumed to include the effects of such a resonating lumped reactor.

The total observable energy stored in such a resonant antenna system, including the energy $|\langle\langle W_m\rangle\rangle_A - \langle\langle W_e\rangle\rangle_A|$ stored in the lumped reactor representing the cavity, is then

$$\langle\langle W_m\rangle\rangle + \langle\langle W_e\rangle\rangle = \langle\langle W_m\rangle\rangle_A + \langle\langle W_e\rangle\rangle_A + |\langle\langle W_m\rangle\rangle_A - \langle\langle W_e\rangle\rangle_A|$$
$$= \begin{cases} 2\langle\langle W_m\rangle\rangle_A, & \langle\langle W_m\rangle\rangle_A > \langle\langle W_e\rangle\rangle_A \\ 2\langle\langle W_e\rangle\rangle_A, & \langle\langle W_m\rangle\rangle_A < \langle\langle W_e\rangle\rangle_A. \end{cases} \tag{5.13}$$

The quality factor (3.33) for the complete antenna system thus reduces to

$$Q = \frac{2\omega\langle\langle W_{me}\rangle\rangle_A}{\langle P\rangle_A}, \tag{5.14}$$

where $\langle\langle W_{me}\rangle\rangle_A$ denotes the greater of the two observable energies (3.38) that are stored in the fields of any planar source and $\langle P\rangle_A$ denotes the time-average power (3.5) that is radiated by the source.

In the particular case of the idealized planar dipole antenna for which detailed calculations were carried out in section 3.7 the source had to be self-resonant, since no exciting cavity was present. The observable stored magnetic and electric energies of the source alone were equal, which meant that no lumped reactor was necessary for resonance.

General expressions for the Q of arbitrary strip and line sources will now be derived. They will be based on the formulas in Chapter 3 for the radiative power and the observable reactive powers of planar sources and on this assumption that the overall effect of the cavity can be represented by a single lumped reactor.

(i) *E-plane strip sources*

In the asymptotic case of E-plane strip sources the pattern space factor was found to collapse to a delta-like function about the n_x axis (Fig. 5.2a). Hence the observable inductive power (3.38*a*) approaches zero because of the vanishing factor n_y^2 in the numerator of the integrand. The total observable stored energy then becomes purely electric. This characteristic property of all E-plane strip sources is shared also by all circularly symmetric radial sources, for which the observable inductive power (3.40*a*) was also seen to be precisely zero, although for a quite different reason. In both cases, then, the quality factor (5.14) is always determined by the observable stored *electric* energy $\langle\langle W_e\rangle\rangle_A$.

The asymptotic form of the radiative and reactive power integrals as the width approaches infinity can be determined by first treating the complex power integral (3.4) as a whole and then separating it into its real (radiative and imaginary (reactive) parts:†

$$\frac{k^2}{(2\pi)^2 2Z_0}\int\limits_{-\infty}^{\infty}\int\limits_{-\infty}^{\infty}\frac{(1-n_y^2)|F_x(kn_x, kn_y, k)|^2}{\{(1-n_x^2-n_y^2)^{\frac{1}{2}}\}^*}\,dn_x\,dn_y$$

$$\sim\frac{k^2}{(2\pi)^2 2Z_0}\int\limits_{-\infty}^{\infty}\frac{|F_{xx}(c_x n_x)|^2}{\{(1-n_x^2)^{\frac{1}{2}}\}^*}\left\{\int\limits_{-\infty}^{\infty}|W_{\alpha_{\|}}(c_y n_y)|^2\,dn_y\right\}dn_x$$

$$= C_1\int\limits_{-\infty}^{\infty}\frac{|F_{xx}(c_x n_x)|^2}{\{(1-n_x^2)^{\frac{1}{2}}\}^*}\,dn_x$$

$$= C_1\left[\int\limits_{-1}^{1}\frac{|F_{xx}(c_x n_x)|^2}{(1-n_x^2)^{\frac{1}{2}}}\,dn_x - i\int\limits_{-\infty;1}^{-1;\infty}\frac{|F_{xx}(c_x n_x)|^2}{(n_x^2-1)^{\frac{1}{2}}}\,dn_x\right]. \tag{5.15}$$

All of the constant factors not associated with $F_{xx}(c_x n_x)$ have been lumped together in C_1, including the integral of the square of (5.3). The real part of (5.15) represents the radiated power and the imaginary part represents the observable capacitive power,

$$\langle P\rangle_A \sim C_1\int\limits_{-1}^{1}\frac{|F_{xx}(c_x n_x)|^2}{(1-n_x^2)^{\frac{1}{2}}}\,dn_x, \tag{5.16}$$

$$2\omega\langle\langle W_e\rangle\rangle_A \sim C_1\int\limits_{-\infty;1}^{-1;\infty}\frac{|F_{xx}(c_x n_x)|^2}{(n_x^2-1)^{\frac{1}{2}}}\,dn_x. \tag{5.17}$$

It is not without interest that these formulas for radiated power and observable capacitive power coincide exactly with the real and imaginary parts of the complex power obtained by Woodward and Lawson‡ in their analysis of the complement to E-plane strip sources. But Woodward and Lawson interpreted their imaginary part as determining only the difference between the stored magnetic and electric energies. It can now be reinterpreted, more specifically, as determining the observable stored magnetic energy alone, since the observable stored electric energy is exactly zero (remember that the stored magnetic and electric energies interchange for complementary apertures).

From the ratio of (5.17) to (5.16) the quality factor Q of any E-plane

† By the notation $\int_{-\infty;1}^{-1;\infty}$ is meant the integral over the broken interval $(-\infty, -1)$ and $(1, \infty)$. It is equal to the difference between the integrals on $(-\infty, \infty)$ and $(-1, 1)$.

‡ P. M. Woodward and J. D. Lawson, *loc. cit.*, eqn (31). The boundary condition at infinity requires that the negative root (2.10) be taken in the invisible region, hence it appears that the wrong root was taken on the right-hand side of their eqn (33).

strip source is seen to be

$$Q = \frac{\int\limits_{-\infty;1}^{-1;\infty} \frac{|F_{xx}(c_x n_x)|^2}{(n_x^2-1)^{\frac{1}{2}}}\,dn_x}{\int\limits_{-1}^{1} \frac{|F_{xx}(c_x n_x)|^2}{(1-n_x^2)^{\frac{1}{2}}}\,dn_x}, \tag{5.18}$$

according to its definition (5.14).

(ii) *H-plane strip sources*

In the asymptotic case of H-plane strip sources the pattern space factor again collapses to a delta-like function, this time about the n_y axis (Fig. 5.2b). Unlike the case of E-plane strip sources, however, neither of the two observable reactive powers in (3.38) can vanish without the source itself vanishing everywhere (because of analyticity of the pattern space factor). The two observable reactive powers differ only by the presence of the factor n_y^2 in the numerator of the integrand of (3.38a), and since that factor is always greater than unity it follows that the observable inductive power will always be greater than the observable capacitive power. Hence the quality factor Q for H-plane strip sources will always be determined by the observable stored *magnetic* energy $\langle\langle W_m\rangle\rangle_A$.

The asymptotic form of the radiative and reactive power integrals as the width approaches infinity can again be determined by treating the complex power integral (3.4) as a whole and then separating it into its real and imaginary parts:

$$\frac{k^2}{(2\pi)^2 2Z_0}\int\limits_{-\infty}^{\infty}\int\limits_{-\infty}^{\infty} \frac{(1-n_y^2)\,|F_x(kn_x, kn_y, k)|^2}{\{(1-n_x^2-n_y^2)^{\frac{1}{2}}\}^*}\,dn_x\,dn_y$$

$$\sim \frac{k^2}{(2\pi)^2 2Z_0}\int\limits_{-\infty}^{\infty} \frac{(1-n_y^2)\,|F_{xy}(c_y n_y)|^2}{\{(1-n_y^2)^{\frac{1}{2}}\}^*}\left\{\int\limits_{-\infty}^{\infty} |W_{\alpha_\perp}(c_x n_x)|^2\,dn_x\right\}dn_y$$

$$= C_2\int\limits_{-\infty}^{\infty} \frac{(1-n_y^2)\,|F_{xy}(c_y n_y)|^2}{\{(1-n_y^2)^{\frac{1}{2}}\}^*}\,dn_y$$

$$= C_2\Bigg[\int\limits_{-1}^{1} |F_{xy}(c_y n_y)|^2\,(1-n_y^2)^{\frac{1}{2}}\,dn_y +$$

$$+i\left\{\int\limits_{-\infty;1}^{-1;\infty} \frac{n_y^2\,|F_{xy}(c_y n_y)|^2}{(n_y^2-1)^{\frac{1}{2}}}\,dn_y - \int\limits_{-\infty;1}^{-1;\infty} \frac{|F_{xy}(c_y n_y)|^2}{(n_y^2-1)^{\frac{1}{2}}}\,dn_y\right\}\Bigg], \tag{5.19}$$

where all of the constant factors not associated with $F_{xy}(c_y n_y)$ have been lumped together in C_2. These three terms are identifiable with radiated power, observable inductive power, and observable capacitive power, respectively,

$$\langle P\rangle_A \sim C_2 \int_{-1}^{1} |F_{xy}(c_y n_y)|^2 (1-n_y^2)^{\frac{1}{2}}\, dn_y, \tag{5.20}$$

$$2\omega\langle\langle W_m\rangle\rangle_A \sim C_2 \int_{-\infty;1}^{-1;\infty} \frac{n_y^2 |F_{xy}(c_y n_y)|^2}{(n_y^2-1)^{\frac{1}{2}}}\, dn_y, \tag{5.21}$$

$$2\omega\langle\langle W_e\rangle\rangle_A \sim C_2 \int_{-\infty;1}^{-1;\infty} \frac{|F_{xy}(c_y n_y)|^2}{(n_y^2-1)^{\frac{1}{2}}}\, dn_y. \tag{5.22}$$

The quality factor Q of H-plane strip sources is then

$$Q = \frac{\displaystyle\int_{-\infty;1}^{-1;\infty} \frac{n_y^2 |F_{xy}(c_y n_y)|^2}{(n_y^2-1)^{\frac{1}{2}}}\, dn_y}{\displaystyle\int_{-1}^{1} |F_{xy}(c_y n_y)|^2 (1-n_y^2)^{\frac{1}{2}}\, dn_y}, \tag{5.23}$$

again from the definition (5.14).

(iii) *H-plane line sources*

The remarkable simplicity of the reactive power integrals obtained for the asymptotic case of each of the two strip sources above is lost completely in the asymptotic case of line sources. This difference between strip sources, on the one hand, and line sources, on the other, occurs because the pattern space factor no longer collapses to a delta-like function about one of the principal axes in the (n_x, n_y) plane but, instead, tends to spread out uniformly along the n_x axis. Because of this spreading the asymptotic form of the observable reactive powers must be derived from their general form.

The general form of the observable inductive power (3.38a) is

$$\begin{aligned} 2\omega\langle\langle W_m\rangle\rangle_A = \frac{E_0^2 a^2 b^2 k^2 \Gamma^2(\alpha_\perp+1)}{2^{5-2\alpha_\perp}\pi Z_0}\Biggl(\int_{-1}^{1} n_y^2 |F_{xy}(c_y n_y)|^2 \times \\ \times\left[\int_{(1-n_y^2)^{\frac{1}{2}}}^{\infty} \left\{\frac{J_{\alpha_\perp+\frac{1}{2}}(c_x n_x)}{(c_x n_x)^{\alpha_\perp+\frac{1}{2}}}\right\}^2 \frac{dn_x}{\{n_x^2-(1-n_y^2)\}^{\frac{1}{2}}}\right] dn_y + \\ + \int_{-\infty;1}^{-1;\infty} n_y^2 |F_{xy}(c_y n_y)|^2 \times \\ \times\left[\int_{0}^{\infty} \left\{\frac{J_{\alpha_\perp+\frac{1}{2}}(c_x n_x)}{(c_x n_x)^{\alpha_\perp+\frac{1}{2}}}\right\}^2 \frac{dn_x}{\{n_x^2+(n_y^2-1)\}^{\frac{1}{2}}}\right] dn_y\Biggr), \end{aligned} \tag{5.24}$$

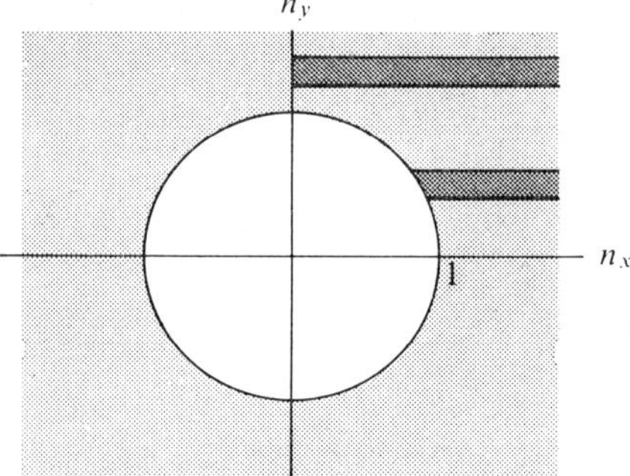

FIG. 5.3. Integration over the invisible region $n_x^2+n_y^2 > 1$ of H-plane line sources.

obtained by integrating over the invisible region $n_x^2+n_y^2 > 1$ in strips as indicated in Fig. 5.3. The inner integrals can be evaluated by replacing the Bessel function by its integral representation on (0, 1). As c_x approaches zero they approach asymptotically the value

$$\int\limits_{(1-n_y^2)^{\frac{1}{2}}}^{\infty} \left\{\frac{J_{\alpha_\perp+\frac{1}{2}}(c_x n_x)}{(c_x n_x)^{\alpha_\perp+\frac{1}{2}}}\right\}^2 \frac{dn_x}{\{n_x^2-(1-n_y^2)\}^{\frac{1}{2}}}$$

$$= -\frac{1}{2^{2\alpha_\perp+1}\Gamma^2(\alpha_\perp+1)} \int\limits_0^1 \int\limits_0^1 \{Y_0(c_x[1-n_y^2]^{\frac{1}{2}}\,|t+t'|)+Y_0(c_x[1-n_y^2]^{\frac{1}{2}}\,|t-t'|)\}\times$$

$$\times(1-t^2)^{\alpha_\perp}(1-t'^2)^{\alpha_\perp}\,dt\,dt'$$

$$\sim -\frac{1}{2^{2\alpha_\perp+1}\Gamma^2(\alpha_\perp+1)}\left[\{C+\ln \tfrac{1}{2}c_x(1-n_y^2)^{\frac{1}{2}}\}\frac{\Gamma^2(\alpha_\perp+1)}{\Gamma^2(\alpha_\perp+\frac{3}{2})}+\right.$$

$$\left.+\frac{2}{\pi}\int\limits_0^1\int\limits_0^1 (1-t^2)^{\alpha_\perp}(1-t'^2)^{\alpha_\perp} \ln|t^2-t'^2|\,dt\,dt'\right], \tag{5.25a}$$

$$\int\limits_0^{\infty} \left\{\frac{J_{\alpha_\perp+\frac{1}{2}}(c_x n_x)}{(c_x n_x)^{\alpha_\perp+\frac{1}{2}}}\right\}^2 \frac{dn_x}{\{n_x^2+(n_y^2-1)\}^{\frac{1}{2}}}$$

$$= \frac{1}{2^{2\alpha_\perp+1}\Gamma^2(\alpha_\perp+1)}\frac{2}{\pi} \int\limits_0^1 \int\limits_0^1 \{K_0(c_x[n_y^2-1]^{\frac{1}{2}}\,|t+t'|)+K_0(c_x[n_y^2-1]^{\frac{1}{2}}\,|t-t'|)\}\times$$

$$\times(1-t^2)^{\alpha_\perp}(1-t'^2)^{\alpha_\perp}\,dt\,dt'$$

$$\sim -\frac{1}{2^{2\alpha_\perp+1}\Gamma^2(\alpha_\perp+1)}\left[\{C+\ln \tfrac{1}{2}c_x(n_y^2-1)^{\frac{1}{2}}\}\frac{\Gamma^2(\alpha_\perp+1)}{\Gamma^2(\alpha_\perp+\frac{3}{2})}+\right.$$

$$\left.+\frac{2}{\pi}\int\limits_0^1\int\limits_0^1 (1-t^2)^{\alpha_\perp}(1-t'^2)^{\alpha_\perp} \ln|t^2-t'^2|\,dt\,dt'\right], \tag{5.25b}$$

where C denotes Euler's constant. Then (5.24) approaches the value

$$2\omega\langle\langle W_m\rangle\rangle_A \sim \frac{E_0^2a^2b^2k^2}{16\pi^2 Z_0}\Bigg[\Bigg\{-\frac{\pi}{4}\frac{\Gamma^2(\alpha_\perp+1)}{\Gamma^2(\alpha_\perp+\frac{3}{2})}\Big(C+\ln\tfrac{1}{2}c_x\Big)-$$

$$-\tfrac{1}{2}\int_0^1\int_0^1(1-t^2)^{\alpha_\perp}(1-t'^2)^{\alpha_\perp}\ln|t^2-t'^2|\,dt\,dt'\Bigg\}\times$$

$$\times\int_{-\infty}^{\infty} n_y^2\,|F_{xy}(c_y n_y)|^2\,dn_y-$$

$$-\frac{\pi}{8}\frac{\Gamma^2(\alpha_\perp+1)}{\Gamma^2(\alpha_\perp+\frac{3}{2})}\int_{-\infty}^{\infty} n_y^2\,|F_{xy}(c_y n_y)|^2\ln|1-n_y^2|\,dn_y\Bigg]. \quad (5.26)$$

The observable capacitive power (3.38*b*) will be precisely the same as (5.26) except that the factor n_y^2 in the integrals will be absent.

The radiative power represented by (3.5) can be evaluated by the same procedure. As c_x approaches zero it approaches asymptotically the value

$$\langle P\rangle_A = \frac{E_0^2a^2b^2k^2}{16\pi^2 Z_0}\frac{\pi}{4}\int_{-1}^{1}(1-n_y^2)\,|F_{xy}(c_y n_y)|^2\times$$

$$\times\Bigg[\int_0^1\int_0^1\{J_0(c_x[1-n_y^2]^{\frac{1}{2}}\,|t+t'|)+$$

$$+J_0(c_x[1-n_y^2]^{\frac{1}{2}}\,|t-t'|)\}(1-t^2)^{\alpha_\perp}(1-t'^2)^{\alpha_\perp}\,dt\,dt'\Bigg]\,dn_y$$

$$\sim\frac{E_0^2a^2b^2k^2}{16\pi^2 Z_0}\frac{\pi^2\Gamma^2(\alpha_\perp+1)}{8\Gamma^2(\alpha_\perp+\frac{3}{2})}\int_{-1}^{1}(1-n_y^2)\,|F_{xy}(c_y n_y)|^2\,dn_y. \quad (5.27)$$

It is now apparent that the value of Q for line sources is complicated in part by its dependence upon the edge exponent $\alpha_\perp$. To simplify matters the exponent will be assumed to have the approximate value $\alpha_\perp = 0$, resulting in a uniform distribution across the vanishingly narrow slit. Then (5.26) and

(5.27) become

$$2\omega\langle\langle W_m\rangle\rangle_A \sim \frac{E_0{}^2a^2b^2k^2}{16\pi^2Z_0}\left\{(\tfrac{3}{2}-C-\ln c_x)\int\limits_{-\infty}^{\infty} n_y{}^2\,|F_{xy}(c_yn_y)|^2\,dn_y - \right.$$

$$\left. -\tfrac{1}{2}\int\limits_{-\infty}^{\infty} n_y{}^2\,|F_{xy}(c_yn_y)|^2 \ln|1-n_y{}^2|\,dn_y\right\}, \tag{5.28}$$

$$\langle P\rangle_A \sim \frac{E_0{}^2a^2b^2k^2}{16\pi^2Z_0}\,\frac{\pi}{2}\int\limits_{-1}^{1}(1-n_y{}^2)\,|F_{xy}(c_yn_y)|^2\,dn_y. \tag{5.29}$$

An important special case was obtained earlier in (3.45) and (3.10), respectively, for the complementary planar dipole antenna, except for the factor $n_y{}^2$ in the integrands of (5.28) since $\langle\langle W_m\rangle\rangle_A$ and $\langle\langle W_e\rangle\rangle_A$ interchange for complementary apertures. In that particular case the distribution along the length of the aperture was taken to be sinusoidal.

As c_x approaches zero the second term in (5.28) is dominated by the logarithmically increasing first term, although rather slowly since logarithmic singularities are very weak. Thus the observable inductive and capacitive powers of any line source with edge exponent $\alpha_\perp = 0$ will increase asymptotically with decreasing c_x as

$$2\omega\langle\langle W_m\rangle\rangle_A \sim \left(\frac{E_0{}^2a^2b^2k^2}{16\pi^2Z_0}\ln\frac{2{\cdot}52}{c_x}\right)\int\limits_{-\infty}^{\infty} n_y{}^2\,|F_{xy}(c_yn_y)|^2\,dn_y, \tag{5.30}$$

$$2\omega\langle\langle W_e\rangle\rangle_A \sim \left(\frac{E_0{}^2a^2b^2k^2}{16\pi^2Z_0}\ln\frac{2{\cdot}52}{c_x}\right)\int\limits_{-\infty}^{\infty} |F_{xy}(c_yn_y)|^2\,dn_y. \tag{5.31}$$

Comparison of the integrals in (5.30) and (5.31) with those of strip sources reveals an essential difference between the energy storage properties of strip and of line sources. Whereas the observable reactive powers of strip sources were seen from (5.17), and again from (5.21) and (5.22), to be determined by just the invisible region alone of the one-dimensional pattern function $F_{xx}(c_xn_x)$ or $F_{xy}(c_yn_y)$, in the case of line sources they are determined by *both* the visible and the invisible regions of $F_{xy}(c_yn_y)$.

This dependence of the observable reactive powers of line sources upon the visible as well as the invisible region of the one-dimensional pattern function adds a further complication to the formulation of Q: the decision as to which of the two reactive powers is the greater must now be based upon an *ad hoc* comparison of the integrals. Thus, the Q of H-plane line sources with edge exponent $\alpha_\perp = 0$ approaches asymptotically the value

$$Q \sim \left(\frac{2}{\pi}\ln\frac{2\cdot 52}{c_x}\right)\frac{\int\limits_{-\infty}^{\infty}\begin{Bmatrix}n_y^2\\1\end{Bmatrix}|F_{xy}(c_y n_y)|^2\,dn_y}{\int\limits_{-1}^{1}(1-n_y^2)\,|F_{xy}(c_y n_y)|^2\,dn_y}, \tag{5.32}$$

from (5.14), where the choice between the two factors n_y^2 and 1 is understood to be the one that results in the larger value of (5.32).

An example is provided by the idealized resonant planar dipole antenna investigated in detail in section 3.7 as a test of the observable stored energies of planar sources. In that particular case (5.32) approaches the asymptotic expression for Q obtained there.

It should be noted that the value (5.32) for the Q of line sources is dependent upon the width ($c_x = \pi a/\lambda$) as well as the length ($c_y = \pi b/\lambda$) of the source, whereas for strip sources the value of Q is completely independent of the width. That the Q of line sources should depend upon width is not really surprising in view of the results obtained for the planar dipole antenna. But it does seem somewhat surprising that this point had not been anticipated earlier in any of the many published allusions to the role that Q was expected to play in the synthesis of line sources.

5.5. Constraining parameter γ_α

All three of the formulas (5.18), (5.23), and (5.32) are exact expressions for the quality factor Q of strip and line sources. Hence, for any preassigned value of Q they represent the constraining condition that must be imposed in order to find an optimum Q-constrained pattern space factor. Like many exact formulations of physical entities, however, they do not fit neatly into the mathematical framework available, namely, that which is provided by the characteristic functions $\{\psi_{\alpha n}(c,\eta)\}$, although they do appear to come surprisingly close. The difficulty lies in the weighting factors contained within the integrals defining Q: the weighting factors are independent of the value of the edge exponent α, whereas those in the double orthogonality integrals (5.12) are wholly dependent upon α. Furthermore there is no systematic relationship among the weighting factors in the expressions for Q of the three different kinds of strip and line sources. The problem now is to attempt to reconcile these differences. To that end a new constraining parameter γ_α will be defined that fits precisely into the mathematical framework provided by the characteristic functions and that, at the same time, bears a simple relationship to the true value of Q, either exactly or to a good approximation.

An investigation by Taylor† on the effects of superdirective line sources

† T. T. Taylor (1955), *loc. cit.*

led him to define a ratio γ that is something like Q,

$$\gamma = \frac{\int\limits_{-\infty}^{\infty} |F_0(c\eta)|^2 \, d\eta}{\int\limits_{-1}^{1} |F_0(c\eta)|^2 \, d\eta}, \tag{5.33}$$

where $F_0(c\eta)$ is the pattern space factor (5.9) for $\alpha = 0$. He showed that pattern space factors with arbitrarily high directivity could be obtained from the pattern of a uniform distribution by shifting some of its zeros from the invisible region into the visible region. One effect of this increase in directivity is to increase the value of the ratio γ, which led him to call γ the superdirectivity ratio. The name is somewhat misleading, however, because the converse need not be true; a large value of γ need not imply a highly directive aperture distribution. The uniform distribution itself, for example, is never considered to be superdirective, and yet its value of γ can be made arbitrarily large simply by shortening the length of the aperture. Hence, γ is clearly not a measure of superdirectivity.

Furthermore γ is not a true measure of Q, either, as is evident by comparison with the expressions developed in section 5.4 for the Q of strip and line sources. But it is at least suggestive of the ratios that appear in the expressions for Q. In the case of strip sources it is somewhat similar to $Q+1$, and in the case of line sources it is proportional (with an arbitrarily large proportionality constant) to a ratio that is somewhat similar to Q. There are still two features missing. One is the weighting factors in the integrals, and the other is the dependence of the pattern space factor upon the edge exponent α. Both can be included, however, either exactly or to a good approximation, by defining the following generalized superdirectivity ratio:

$$\gamma_\alpha^\beta = \frac{\int\limits_{-\infty}^{\infty} |F_\alpha(c\eta)|^2 \, |1-\eta^2|^\beta \, d\eta}{\int\limits_{-1}^{1} |F_\alpha(c\eta)|^2 (1-\eta^2)^\beta \, d\eta}. \tag{5.34}$$

It reduces precisely to Taylor's γ when $\alpha = \beta = 0$. But for certain other values of α and β it has some distinct advantages over Taylor's γ. It can be made exactly equal to $Q+1$ for E-plane strip sources by choosing the weighting exponent β to be $-\frac{1}{2}$. And it gives a good approximation to $Q+1$ for H-plane strip sources by choosing β to be $\frac{1}{2}$. For H-plane line sources it leads to a quantity that is approximately proportional to Q by choosing β to be 1.

The generalization γ_α^β is now too general, however, to fit the mathematical framework provided by the characteristic functions $\{\psi_{\alpha n}(c, \eta)\}$. Whereas γ_α^β will exist as a mathematical entity whenever the edge exponent α and the weighting exponent β in (5.34) are any two real numbers greater than -1, the double orthogonality property (5.12) occurs only when they are equal to each other. In order to be able to make use of this powerful property of double orthogonality it will be necessary to assume that the values chosen for the edge exponent α and for the weighting exponent β are equal to each other. The resulting ratio γ_α^α will be denoted simply by

$$\gamma_\alpha = \frac{\int_{-\infty}^{\infty} |F_\alpha(c\eta)|^2\,|1-\eta^2|^\alpha\,d\eta}{\int_{-1}^{1} |F_\alpha(c\eta)|^2\,(1-\eta^2)^\alpha\,d\eta}. \tag{5.35}$$

This is the *constraining parameter* γ_α that will be used to obtain the optimum synthesis solutions in Chapters 8 and 9.

With γ_α as the constraining parameter instead of Q the question now arises as to whether, and how, the value of α can be chosen so as to insure that the performance of the constrained synthesis solution will be valid physically. This is not a trivial question because the physical validity of the performance represented by the solution now rests entirely upon the validity of the assumption that a single number α can be used to specify both the edge behaviour and the weighting factors simultaneously in such a way that the γ_α-constrained solution will be an acceptable approximation to the true Q-constrained solution. Any questions on the effect of edge behaviour can be dispensed with immediately, however, from the conclusion drawn in section 2.8(iv) that the performance available from any planar source will be completely insensitive to the particular choice of edge behaviour, in spite of the fact that only one choice appears to be realizable physically. This leaves only the question of whether some value of α exists such that the resulting value of γ_α will bear a close relationship to the value of Q. It is a question that takes on somewhat different forms for each of the three different types of strip and line sources, hence each will again be treated separately.

(i) *E-plane strip sources*

A relationship between γ_α and Q for E-plane strip sources can now be written that is simple and exact. Comparing (5.18) with (5.35) it is evident that γ_α is related to Q by

$$Q = \gamma_{-\frac{1}{2}} - 1, \tag{5.36}$$

obtained by choosing the value of $\alpha_\perp$ to be $-\frac{1}{2}$.

It is of interest to note that this value $\alpha_\perp = -\frac{1}{2}$ is the absolute minimum value (2.69*a*) of $\alpha_{\perp}$, that it is the value that would occur for a knife edge, and that the characteristic spheroidal functions for that particular value are the even Mathieu functions $\{Se_n(c, \eta)\}$.

(ii) *H-plane strip sources*

In the case of *H*-plane strip sources there is no value of $\alpha_{\parallel}$ for which an exact relationship exists between γ_{α} and Q. But if the value of $\alpha_{\parallel}$ is chosen to be $+\frac{1}{2}$ one can see from (5.23) that the denominator of (5.35) will be exactly correct and that the numerator of (5.35) accounts exactly for the complex power flow (5.19) through the aperture. On the other hand, the imaginary part of the complex power represents only the difference between the observable inductive and capacitive reactive powers of the unresonated source alone. While the difference is admittedly a quantity that is of great physical interest it fails to endow $\gamma_{\frac{1}{2}}$ with the desired property of being related exactly to Q. It is the sum, not the difference, of the observable reactive powers of the source that determines its Q.

Since the observable inductive power of any *H*-plane strip source always exceeds the observable capacitive power it is evident by comparing (5.23) with (5.35) that the following approximate relationship holds,

$$Q \doteqdot \gamma_{\frac{1}{2}} - 1, \tag{5.37}$$

obtained by choosing the value of $\alpha_{\parallel}$ to be $+\frac{1}{2}$. It will be most accurate whenever the observable inductive power far exceeds the observable capacitive power, which will occur whenever the largest invisible lobes are located far from the visible region.

Again it is of interest to note that this value $\alpha_{\parallel} = +\frac{1}{2}$ is the absolute minimum value (2.69*b*) of $\alpha_{\parallel}$, that it is the value that would occur for a knife edge, and that the characteristic spheroidal functions for that particular value are the odd Mathieu functions $\{So_{n+1}(c, \eta)\}$ divided by $(1-\eta^2)^{\frac{1}{2}}$.

(iii) *H-plane line sources*

For *H*-plane line sources the weighting factor in the integral in the donominator of (5.35) will be exactly correct only for $\alpha_{\parallel} - 1$, from expression (5.32) for Q, whereas the weighting factor in the numerator can be exactly correct only for $\alpha_{\parallel} = 0$ and then only when the observable capacitive power predominates. Thus, as in the case of *H*-plane strip sources there is no value of α for which γ_{α} is related exactly to Q. But again there is an approximate relationship, the approximation being best for $\alpha_{\parallel} = 1$ if the pattern is concentrated either about the broadside direction $n_y = 0$ (small Q) or far away from the visible region (large Q),

$$Q \mathrel{\dot{\approx}} \left(\frac{2}{\pi} \ln \frac{2{\cdot}52}{c_x}\right) \gamma_1(c_y). \tag{5.38}$$

The dependence upon width (in c_x) and upon length (in c_y) of the source is shown here explicitly.

The characteristic spheroidal functions for H-plane strip sources are then the prolate spheroidal wave functions of order unity $\{S_{1n}(c, \eta)\}$ divided by $(1-\eta^2)^{\frac{1}{2}}$.

5.6. Absolute lower bound on γ_α

An important property of γ_α, and hence of the value of Q of strip and line sources, is that it possesses an absolute lower bound below which no strip or line source can exist. This comes from an extremal property of the spheroidal functions. The reciprocal of γ_α is a measure of the extent to which the aperture-limited function $F_\alpha(c\eta)$ is concentrated within the visible region $-1 < \eta < 1$. The extrema of this measure of concentration are shown in Appendix 3.5 to be the reciprocal of the infinite set of characteristic numbers $\{\gamma_{\alpha n}(c)\}$ of the spheroidal functions of order α. From this it follows that the lower bound on γ_α is the smallest of the characteristic numbers $\{\gamma_{\alpha n}(c)\}$, which is known to be the one for $n = 0$, and that any aperture-limited pattern space factor $F(c\eta)$ whose value of γ_α coincides with this lower bound $\gamma_{\alpha 0}(c)$ must be precisely proportional to the corresponding spheroidal function $\psi_{\alpha 0}(c, \eta)$. It is this connection between γ_α and the characteristic numbers of the spheroidal functions that led the author originally to the choice of notation $\gamma_{\alpha n}(c)$ for these characteristic numbers. And it is this uniqueness of the pattern space factor for $\gamma_\alpha = \gamma_{\alpha 0}(c)$ that will be seen to determine the correct extremum for the maximum directivity solution in Chapter 8.

It is concluded, therefore, that whenever the parameter γ_α is imposed as a constraint upon the solution to a strip or a line source synthesis problem the value assigned to it must always be chosen such that its value is not less than the smallest characteristic number $\gamma_{\alpha 0}(c)$ of the spheroidal functions; i.e. such that

$$\gamma_\alpha \geq \gamma_{\alpha 0}(c), \tag{5.39}$$

since no solution at all is possible for values of γ_α that are smaller than $\gamma_{\alpha 0}(c)$. Some numerical values of the reciprocal of this absolute lower bound are included in Fig. A3.4 and in Table A3.2 of Appendix 3.

CHAPTER 6

SAMPLING SYNTHESIS

SERIES representations for aperture-limited pattern functions usually fall into one of two different categories. In one category each coefficient in the series can be obtained from the pattern as a whole by means of an orthogonality property of the basis functions. The spheroidal functions that were introduced in section 5.3 and that will be used to obtain the synthesis solutions in Chapters 8 and 9 are an example of basis functions in this first category. In the other category each coefficient is obtained not from the pattern as a whole but from just one, or at most two, sample values of the pattern function. The resulting basis functions are the sampling functions that will be described here and that will be used to obtain the synthesis solution for maximum resolution in Chapter 7.

Whenever the sampling points are equally spaced it will be seen that the sampling series has the remarkable property of being an orthogonal series as well. In such cases, of which there are only two, the coefficients of the series can be obtained either from discrete values of the pattern function at the sampling points, by virtue of the sampling property of the basis functions, or from the pattern function as a whole, by virtue of the orthogonality property. But even when the sampling points are not equally spaced it will be seen from the general theory of sampling synthesis to be developed here that the sampling functions always form one of a pair of biorthogonal sets. Thus, each coefficient in any sampling series could still be determined equally well from the pattern as a whole by means of this biorthogonality property.

6.1. Woodward synthesis

The theory of sampling synthesis began with Woodward's synthesis method.† Woodward synthesis arose from two observations. One is the fact that the zeros of the aperture-limited pattern $(\sin u)/u$ of a uniform source all lie at integral multiples of π. The other is the fact that the pattern can be shifted in u without distortion by introducing a linear phase variation along the length of the source. By choosing the total phase variation along the normalized length $-1 < t < 1$ to be $2n\pi$ radians the $(\sin u)/u$ pattern will be shifted by an amount $n\pi$, for which its zeros will again all lie at integral multiples of π. Only the nth function can have a nonzero value at the point $u = n\pi$. Any given non-aperture-limited pattern can then be

† P. M. Woodward. A method of calculating the field over a plane aperture required to produce a given polar diagram. *J. Instn. elect. Engrs.* Part IIIA, **93**, pp. 1554–8, 1946.

approximated over the visible region by a linear combination of $\{\sin(u-n\pi)/(u-n\pi)\}$ functions, where the coefficient of the nth function is simply the sample value of the given pattern at the point $u = n\pi$. The resulting pattern approximation will be an aperture-limited function that is best in the sense that it coincides with the given non-aperture-limited pattern at the sampling points $\{n\pi\}$. It is exactly equivalent mathematically to the sampled-data representation of band-limited signals that was proposed two years later by Shannon† and that is usually referred to as Shannon's sampling theorem.

Woodward synthesis can be formulated more precisely as follows. Let the principal-plane pattern space factor (5.4) of a strip or line source of length l be represented in terms of the aperture distribution $A(t)$ sought by

$$F(u) = \int_{-1}^{1} A(t)e^{iut}\,dt, \tag{6.1}$$

where u denotes $(\pi l/\lambda)\sin\theta$, and let $A(t)$ be represented by a sum of uniform travelling waves with a linear phase change of $2n\pi$ radians over the length of the aperture,

$$A(t) = \tfrac{1}{2}\sum_{n=-N}^{N} b_n e^{-in\pi t}, \tag{6.2}$$

where N is the largest whole number of wavelengths contained within the length of the aperture. Then the pattern space factor, obtained by integrating the series term by term, is

$$F(u) = \sum_{n=-N}^{N} b_n \frac{\sin(u-n\pi)}{u-n\pi}. \tag{6.3}$$

It can be made to coincide exactly with any given function $\hat{F}(u)$ at $2N+1$ equally spaced sampling points $u = 0, \pm\pi, \pm 2\pi, \ldots, \pm N\pi$ over the visible region by choosing each coefficient b_n to have the sample value

$$b_n = \hat{F}(n\pi). \tag{6.4}$$

As an example of Woodward synthesis consider the case of an air-to-ground search radar antenna whose E-plane source is tilted at an angle of 45°, as in Fig. 6.1. The objective is to make the field strength at the gound as nearly constant as possible over a given angular interval and zero outside of that interval. Since field strength varies inversely with radial distance the radiation pattern desired is a cosecant function of direction,

$$\hat{F}(u) = \begin{cases} \dfrac{\sqrt{2}}{2}\operatorname{cosec}\left(\arcsin\dfrac{u}{\pi l/\lambda}+\dfrac{\pi}{4}\right), & \left|\dfrac{u}{\pi l/\lambda}\right| < 0{\cdot}5 \\ 0 & , \quad \left|\dfrac{u}{\pi l/\lambda}\right| > 0{\cdot}5, \end{cases} \tag{6.5}$$

† C. E. Shannon, *loc. cit.*, Theorem 13.

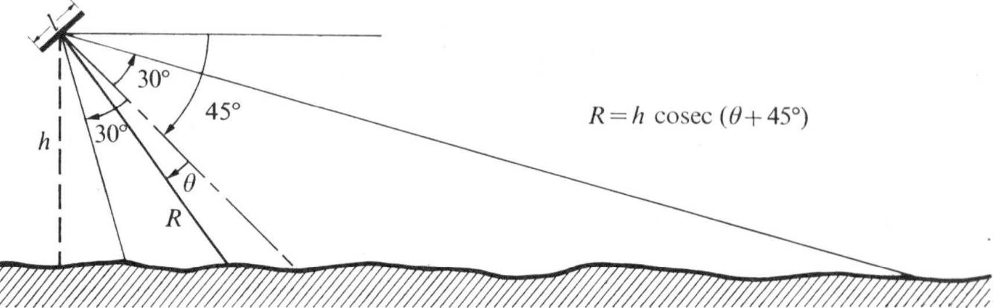

FIG. 6.1. E-plane source of length l at height h above ground, tilted at an angle of 45°.

where the angular interval of constant field strength at the ground is taken here to be $-30° < \theta < 30°$. This is clearly not an aperture-limited function since it is not even analytic. Hence it can only be approximated. For an aperture of length 5λ the Woodward approximation to (6.5) is determined solely by samples located at the five points $u = 0$, $\pm\pi$, and $\pm 2\pi$, all others being zero. The given function $\hat{F}(u)$ and its Woodward approximation (6.3) is shown in Fig. 6.2, together with the magnitude and phase of the source (6.2) required to produce it. From the relative smallness of the Woodward approximation in the invisible region $|u| > \pi l/\lambda$ (shown by the broken line) it is evident that the Q of such an antenna will not be excessive.

The Woodward approximation is exact at only the sampling points. Hence it is evident that any improvement in accuracy can be obtained only by increasing the number of sampling points. Since the number of points is limited to at most $(2l/\lambda)+1$, however, any improvement in accuracy can be obtained only by lengthening the aperture.

An alternative synthesis method that would increase the accuracy of the pattern approximation without increasing the length of the aperture was proposed later by Woodward and Lawson.† They observed that a source in

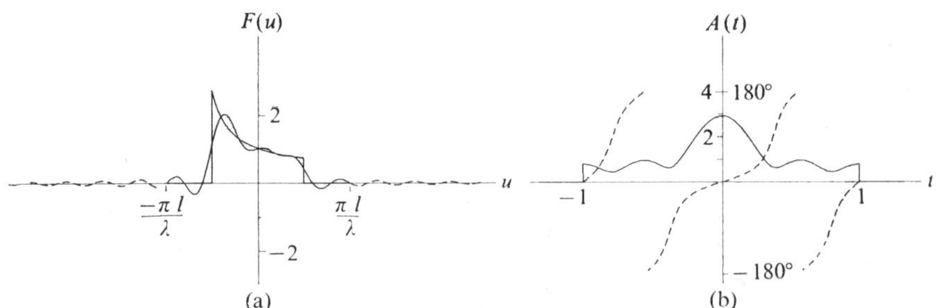

FIG. 6.2. Woodward synthesis for the cosecant pattern (6.5) in the case of an E-plane source of length $l = 5\lambda$: (a) pattern approximation, and (b) magnitude (solid) and phase (broken) of the source required.

† P. M. Woodward and J. D. Lawson, *loc. cit.*, p. 365.

an aperture of any given length can be found for which the number of directions at which the value of the radiation pattern can be assigned need not be limited to $(2l/\lambda)+1$ but can be any number whatever. The method consists simply of solving a set of simultaneous equations for as many unknown coefficients as there are directions in which values of the radiation pattern have been assigned. Specifically, if the pattern approximation $F(u)$ is forced to coincide with the given function $\hat{F}(u)$ at any $2N+1$ points $\{u_m\}$ in the interval $(-\pi l/\lambda, \pi l/\lambda)$ for N arbitrarily large, then, from (6.3),

$$\hat{F}(u_m) = \sum_{n=-N}^{N} a_n \frac{\sin(u_m - n\pi)}{u_m - n\pi}, \qquad m = 0, \pm 1, \pm 2, \dots, \pm N. \tag{6.6}$$

By solving this set of $2N+1$ equations for the $2N+1$ unknown a_ns an aperture-limited pattern function (6.3) and its source (6.2) can be constructed that will coincide exactly with the given function $\hat{F}(u)$ at any $2N+1$ points, where N can now be made as large as desired.

The trouble with this synthesis method, as Woodward and Lawson pointed out quite clearly, is that the source would soon become so highly reactive that it would become unrealizable physically. All of the component pattern functions for $|n|$ greater than l/λ would have their main beams in the invisible region. Since their coefficients would have to become very large for their sidelobes to make the contribution required in the visible region, it is evident that their total contribution to the complex power at the aperture would become overwhelmingly reactive.

A far more practical method for improving the accuracy of the Woodward approximation in the case of real-valued functions $\hat{F}(u)$ was proposed recently by Stutzman.† It is based on the observation that the original Woodward approximation $F(u)$ always oscillates about $\hat{F}(u)$, resulting in an error pattern $F(u)-\hat{F}(u)$ that passes through zero at integral multiples of π (the sampling points) and that becomes alternately positive and negative in the intervals between the sampling points. The error will usually be greatest at some point that is roughly half way between adjoining pairs of sampling points. Hence, a good approximation to the error pattern alone can be obtained in the form of a sampling series of functions obtained from $(\sin u)/u$ by shifting it in u by half-integral instead of integral multiples of π. A substantial improvement in pattern accuracy can then be obtained simply by subtracting this sampling series approximation to the error pattern from the Woodward approximation to the original pattern desired. The reduced error will then be of second order. It can be reduced still further by subtracting successive iterations of the error pattern obtained by sampling at points that alternate between half-integral and integral multiples of π. Since the main

† W. L. Stutzman. Synthesis of shaped-beam radiation patterns using the iterative sampling method. *IEEE Trans. Antennas Propagat.* **AP-19**, pp. 36–41, January 1971.

lobe of each of the sampling functions lies within the visible region it is evident that the resulting improvement in accuracy will involve little, if any, increase in the Q of the antenna.

6.2. Taylor's extension of Woodward synthesis

One of the most useful forms of sampling synthesis is an extension of Woodward synthesis by Taylor.† By converting the Woodward series (6.3) into an equivalent product Taylor showed that any Woodward pattern could be described entirely in terms of the location of the zeros of a polynomial in the complex u-plane. This opened up a whole new approach to the synthesis problem, namely, the design of aperture-limited pattern functions by choosing the location of their zeros. The first important example was Taylor's own design of the narrow-beam pattern that will be described later in section 7.1. Instead of taking the desired pattern to be some non-aperture-limited function and then sampling it to obtain an aperture-limited approximation, as Woodward did, Taylor recognized that a closer aperture-limited approximation can frequently be constructed directly by judiciously choosing the position of its central zeros. Once the zeros have been positioned in such a way as to give an acceptable pattern approximation the source required to produce it can then be obtained exactly by means of Woodward sampling of the resulting aperture-limited pattern function.

Taylor's extension of Woodward synthesis was obtained as follows. He converted the Woodward series (6.3) into an equivalent product by first separating out the common factor $\sin u$ and then combining the sum of fractions into a single fraction,

$$F(u) = \sin u \sum_{n=-N}^{N} \frac{(-1)^n b_n}{u-n\pi} = \frac{\sin u}{u} \frac{c_0+c_1u+c_2u^2+\ldots+c_{2N}u^{2N}}{\prod_{n=1}^{N}(u-n\pi)(u+n\pi)}. \tag{6.7}$$

If the coefficient c_{2N} in the polynomial in the numerator is not exactly zero (it would be exactly zero for all odd functions, for example) the polynomial can be factored into a product of $2N$ binomials with zeros $\{u_n\}$,

$$F(u) = \frac{\sin u}{u} \frac{c_{2N}\prod_{n=1}^{2N}(u-u_n)}{\prod_{n=1}^{N}(u-n\pi)(u+n\pi)}. \tag{6.8}$$

† T. T. Taylor. *An antenna pattern synthesis method with discussion of energy storage considerations.* Presented at the joint meeting of the International Scientific Radio Union and IRE Professional Group on Antennas and Propagation, National Bureau of Standards, Washington, D.C., 18 April 1951.

This is clearly equivalent to the Woodward series but it is capable of quite a different interpretation. The $2N$ factors in the denominator of the right-hand ratio have zeros that coincide precisely with the $2N$ central zeros of the function $(\sin u)/u$, the effect of which is to exchange those central zeros for the $2N$ zeros of the polynomial in the numerator. Thus, Taylor's representation (6.8) for the Woodward pattern can be interpreted simply as a modification of the pattern function $(\sin u)/u$ in which each of its $2N$ central zeros has been moved to any arbitrary point u_n in the complex u-plane. A possible configuration for the zeros of the modified pattern is illustrated in Fig. 6.3.

An important feature of this representation is the fact that the new location chosen for the $2N$ central zeros need not be restricted to the real axis. This is

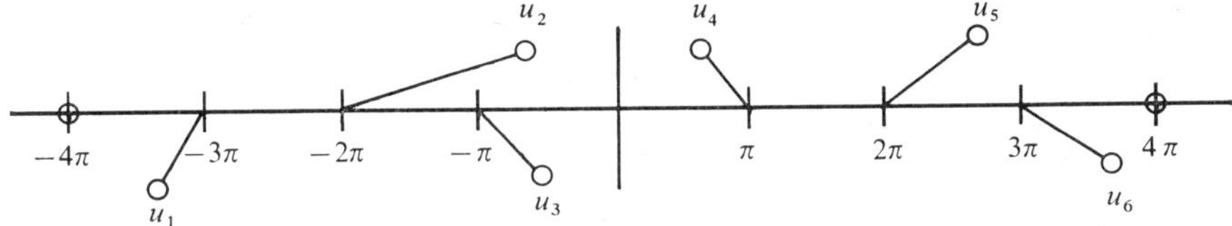

FIG. 6.3. Displacement of the $2N$ central zeros of $(\sin u)/u$ to $2N$ arbitrarily chosen points in the complex u-plane.

particularly important for those pattern-shaping problems in which the phase of the pattern function is of no physical interest, as is frequently the case in practice. In such cases it is usually assumed, either consciously or unconsciously, that the pattern function $F(u)$ must be real-valued for real values of u. But this is an artificial restriction that requires either that the zeros all be real or that any off-axis zeros appear in conjugate pairs. Such a restriction imposes an unnecessary constraint upon the accuracy with which a given pattern-magnitude function can be approximated.

With the Woodward pattern recast into Taylor's form (6.8) the optimum pattern-approximation problem now becomes one of finding the zeros of a polynomial of degree $2N$ that results in the best fit to the given pattern function or, alternatively, to just the magnitude of the given pattern when its phase is of no interest. Thus, all of the well-known techniques for synthesizing polynomials can be brought into play. This represents a considerable and important extension of Woodward's original method.

The number N can be any positive integer, in principle. But if N is chosen greater than the largest whole number of wavelengths contained within the length of the source the synthesis process should be approached with extreme caution because it involves movement of zeros that lie within the invisible region. Movement of invisible zeros can create large invisible lobes, which would require large amounts of reactive power to sustain them and hence could result in a source that is difficult to realize physically.

After an acceptable pattern approximation has been found, by whatever method may be used to construct the approximating polynomial, the source required to produce it will again be given exactly by the Woodward source series (6.2). Here the coefficients $\{b_n\}$ are just the sample values of the aperture-limited approximation (6.8) at the sampling points $\{n\pi\}$. The source will be unique if the phase as well as the magnitude of the given pattern function is specified. Otherwise it will not. Any approximation to just the pattern magnitude alone will remain unchanged whenever a complex zero u_n is replaced by its complex conjugate u_n^*, which means that for $2N$ different complex zeros there will be 2^{2N} different sources whose pattern-magnitude functions will all be identical.

Taylor's extension of Woodward synthesis has been used by Hyneman and Johnson† to develop a practical method for approximating any given pattern-magnitude function. Their method provides a technique for relocating the $2N$ central zeros at complex values of $\{u_n\}$ in a way that gives direct control over the pattern design and that leads to highly accurate beam shaping. The resulting pattern-magnitude approximation gives nearly equal-percentage ripple, the magnitude of the ripple being determined by the relative displacement of the zeros of the pattern function from the real axis.

6.3. Periodic sampling synthesis[13]

A peculiarity of the Woodward distribution (6.2) is that its edge behaviour is different for the even part than for the odd part. The form of the edge behaviour is obscured somewhat by the fact that each of the complex exponential functions $\{e^{-in\pi t}\}$ is a composite of an even and an odd function. But when the distribution is separated into its even and odd parts,

$$A_e(t) = \tfrac{1}{2} \sum_{n=-N}^{N} b_n \cos n\pi t, \tag{6.9a}$$

$$A_o(t) = -i\,\tfrac{1}{2} \sum_{n=-N}^{N} b_n \sin n\pi t, \tag{6.9b}$$

the edge behaviour of each part alone is clear. Each of the even basis functions $\{\frac{1}{2}\cos n\pi t\}$ terminates in a pedestal at the edges $t = \pm 1$, hence the edge exponent α for the even part (6.9a) will have the value $\alpha = 0$, in general. Each of the odd basis functions $\{-i\frac{1}{2}\sin n\pi t\}$, on the other hand, approaches zero linearly at $t = \pm 1$, which means that the edge exponent for the odd part (6.9b) will have the quite different value $\alpha = 1$. Thus, one is led to the curious conclusion that the edge behaviour of the Woodward distribution is a hybrid combination of a pedestal ($\alpha = 0$) for the even part and linear ($\alpha = 1$) for the odd part.

† R. F. Hyneman and R. M. Johnson. A technique for the synthesis of shaped-beam radiation patterns with approximately equal-percentage ripple. *IEEE Trans. Antennas Propagat.* **AP-15**, pp. 736–43, November 1967.

Because of this difference in edge behaviour the Woodward sampling series (6.3) is totally incapable of representing an arbitrary aperture-limited function of order zero or one, even when the number of terms is allowed to become infinite. What is needed is a new set of odd sampling functions for $\alpha = 0$ and a new set of even sampling functions for $\alpha = 1$. These will now be constructed. Together they provide a set of sampling functions that are capable of synthesizing the most general source possible for an edge exponent of either $\alpha = 0$ or 1. The sampling points for the even and the odd parts separately are spaced periodically at intervals of π. Hence this generalization of Woodward synthesis will be referred to here as *periodic sampling synthesis*.

Woodward interpreted his aperture basis functions $\{e^{-in\pi t}\}$ as representing uniform travelling waves. The corresponding physical interpretation of the even and odd parts $\{\cos n\pi t\}$ and $\{-i \sin n\pi t\}$, respectively, is that of standing waves; i.e. of pairs of uniform travelling waves moving in opposite directions. Each pair of travelling waves add together in such a way that they either reinforce or cancel each other at the edges to produce the pedestal or linear form of edge behaviour described by $\alpha = 0$ or 1, respectively. This suggests that the way to construct the missing odd functions for $\alpha = 0$, and the missing even functions for $\alpha = 1$, is to form pairs of uniform travelling waves that again reinforce or cancel, respectively, at the edges.

For $\alpha = 0$ the appropriate pairing of waves is

$$A_e(t) = \tfrac{1}{2} \sum_{\substack{p=-\infty \\ p \text{ even}}}^{\infty} a_p e^{-i(p\pi/2)t} = \sum_{n=0}^{\infty} a_{2n} \frac{\epsilon_n}{2} \cos n\pi t, \tag{6.10a}$$

$$A_o(t) = \tfrac{1}{2} \sum_{\substack{p=-\infty \\ p \text{ odd}}}^{\infty} a_p e^{-i(p\pi/2)t} = -i \sum_{n=0}^{\infty} a_{2n+1} \sin(n+\tfrac{1}{2})\pi t, \tag{6.10b}$$

in which each wave of amplitude $\tfrac{1}{2}a_{-p}$ moving in one direction has been paired with the wave of amplitude $\tfrac{1}{2}a_p$ moving in the opposite direction by choosing them to be related by

$$a_{-p} = (-1)^p a_p, \tag{6.11}$$

where ϵ_n denotes Neumann's number. The even and odd parts of the pattern space factor produced by (6.10) are then, from (6.1),

$$F_e(u) = \sum_{\substack{p=-\infty \\ p \text{ even}}}^{\infty} a_p \frac{\sin(u-\tfrac{1}{2}p\pi)}{u-\tfrac{1}{2}p\pi} = \sum_{n=0}^{\infty} a_{2n} \frac{\epsilon_n}{2} \frac{(-1)^n 2u \sin u}{u^2-\{n\pi\}^2}, \tag{6.12a}$$

$$F_o(u) = \sum_{\substack{p=-\infty \\ p \text{ odd}}}^{\infty} a_p \frac{\sin(u-\tfrac{1}{2}p\pi)}{u-\tfrac{1}{2}p\pi} = \sum_{n=0}^{\infty} a_{2n+1} \frac{(-1)^{n+1} 2u \cos u}{u^2-\{(n+\tfrac{1}{2})\pi\}^2}. \tag{6.12b}$$

For $\alpha = 1$ the appropriate pairing of waves is

$$A_e(t) = \tfrac{1}{2} \sum_{\substack{p=-\infty \\ p \text{ odd}}}^{\infty} a_p e^{-i(p\pi/2)t} = \sum_{n=0}^{\infty} a_{2n+1} \cos(n+\tfrac{1}{2})\pi t, \tag{6.13a}$$

$$A_o(t) = \tfrac{1}{2} \sum_{\substack{p=-\infty \\ p \text{ even}}}^{\infty} a_p e^{-i(p\pi/2)t} = -i \sum_{n=0}^{\infty} a_{2n} \sin n\pi t, \tag{6.13b}$$

in which a_{-p} is related to a_p by

$$a_{-p} = -(-1)^p a_p. \tag{6.14}$$

The even and odd parts of the pattern space factor produced by (6.13) are seen to be

$$F_e(u) = \sum_{\substack{p=-\infty \\ p \text{ odd}}}^{\infty} a_p \frac{\sin(u-\frac{1}{2}p\pi)}{u-\frac{1}{2}p\pi} = \sum_{n=0}^{\infty} a_{2n+1} \frac{(-1)^{n+1}(2n+1)\pi \cos u}{u^2-\{(n+\frac{1}{2})\pi\}^2}, \tag{6.15a}$$

$$F_o(u) = \sum_{\substack{p=-\infty \\ p \text{ even}}}^{\infty} a_p \frac{\sin(u-\frac{1}{2}p\pi)}{u-\frac{1}{2}p\pi} = \sum_{n=0}^{\infty} a_{2n} \frac{(-1)^n 2n\pi \sin u}{u^2-\{n\pi\}^2}. \tag{6.15b}$$

Combining the even and odd parts gives a simple and exact description of the source and its pattern space factor for either $\alpha = 0$ or 1,

$$A(t) = \tfrac{1}{2} \sum_{p=-\infty}^{\infty} a_p e^{-i(p\pi/2)t} = \sum_{p=0}^{\infty} a_p \phi_p(t), \tag{6.16}$$

$$F(u) = \sum_{p=-\infty}^{\infty} a_p \frac{\sin(u-\frac{1}{2}p\pi)}{u-\frac{1}{2}p\pi} = \sum_{p=0}^{\infty} a_p \Phi_p(u), \tag{6.17}$$

in which the aperture basis functions $\{\phi_p(t)\}$ for $\alpha = 0$ are

$$\phi_{2n}(t) = \tfrac{1}{2}\epsilon_n \cos n\pi t, \qquad \phi_{2n+1}(t) = -i \sin(n+\tfrac{1}{2})\pi t, \tag{6.18}$$

and those for $\alpha = 1$ are

$$\phi_{2n}(t) = -i \sin n\pi t, \qquad \phi_{2n+1}(t) = \cos(n+\tfrac{1}{2})\pi t, \tag{6.19}$$

while the corresponding pattern basis functions $\{\Phi_p(u)\}$, defined by

$$\Phi_p(u) = \int_{-1}^{1} \phi_p(t) e^{iut}\, dt, \tag{6.20}$$

for $\alpha = 0$ are

$$\Phi_{2n}(u) = \frac{\epsilon_n}{2} \frac{(-1)^n 2u \sin u}{u^2-\{n\pi\}^2}, \qquad \Phi_{2n+1}(u) = \frac{(-1)^{n+1} 2u \cos u}{u^2-\{(n+\frac{1}{2})\pi\}^2}, \tag{6.21}$$

and for $\alpha = 1$ are

$$\Phi_{2n}(u) = \frac{(-1)^n 2n\pi \sin u}{u^2-\{n\pi\}^2}, \qquad \Phi_{2n+1}(u) = \frac{(-1)^{n+1}(2n+1)\pi \cos u}{u^2-\{(n+\frac{1}{2})\pi\}^2}. \tag{6.22}$$

Each of the pattern basis functions $\{\Phi_{2n}(u)\}$, or $\{\Phi_{2n+1}(u)\}$, has an infinite number of equally spaced zeros that appear at all but two of the even, or odd, multiples of $\frac{1}{2}\pi$, respectively. This is illustrated in Fig. 6.4 for $n = 2$. The two exceptional points are $u = \pm n\pi$, or $\pm(n+\frac{1}{2})\pi$, at which the zeros have been removed. At those two points each even function has the values $+1$, $+1$ and each odd function has the values -1, $+1$. Thus it is seen that each of the pattern basis functions is a sampling function that samples either the even or the odd part, but not both, of the pattern space factor at that particular pair of points.

The set of aperture basis functions $\{\phi_p(t)\}$ is complete with respect to all functions that are square-integrable on $(-1, 1)$. This comes from the theory of Fourier series, where it is known that each set of even basis functions $\{\cos n\pi t\}$ and $\{\cos(n+\frac{1}{2})\pi t\}$ is complete with respect to all even square-integrable functions on $(-1, 1)$ and that each set of odd basis functions $\{\sin n\pi t\}$ and $\{\sin(n+\frac{1}{2})\pi t\}$ is complete with respect to all odd square-integrable functions on $(-1, 1)$. From this it will be shown in section 6.4 that the set of pattern sampling functions $\{\Phi_p(u)\}$ for $\alpha = 0$ or 1 is complete with respect to all aperture-limited functions.

The aperture basis functions $\{\phi_p(t)\}$ are all orthogonal on $(-1, 1)$,

$$\int_{-1}^{1} \phi_p(t)\phi_q^*(t)\,dt = \tfrac{1}{2}\epsilon_p\,\delta_{pq}, \tag{6.23}$$

as can be verified by direct calculation. From this it follows that the pattern sampling functions $\{\Phi_p(u)\}$ are orthogonal on $(-\infty, \infty)$,

$$\int_{-\infty}^{\infty} \Phi_p(u)\Phi_q^*(u)\,du = \int_{-1}^{1}\int_{-1}^{1} \phi_p(t)\phi_q^*(t')\left\{\int_{-\infty}^{\infty} e^{iu(t-t')}\,du\right\} dt\,dt'$$

$$= 2\pi \int_{-1}^{1} \phi_p(t)\phi_q^*(t)\,dt = \pi\epsilon_p\,\delta_{pq}, \tag{6.24}$$

by interchange of the order of integration and recognizing that the integral on u is 2π times the delta function.

It may be of interest to note in passing that the sampling functions $\{\Phi_p(u)\}$ for $\alpha = 1$ are a special case of spheroidal functions of order $\alpha = 1$, and that the orthogonality relations (6.23) and (6.24) constitute a double orthogonality property of these spheroidal functions that is quite different from any known before. This is demonstrated in Appendix 5.

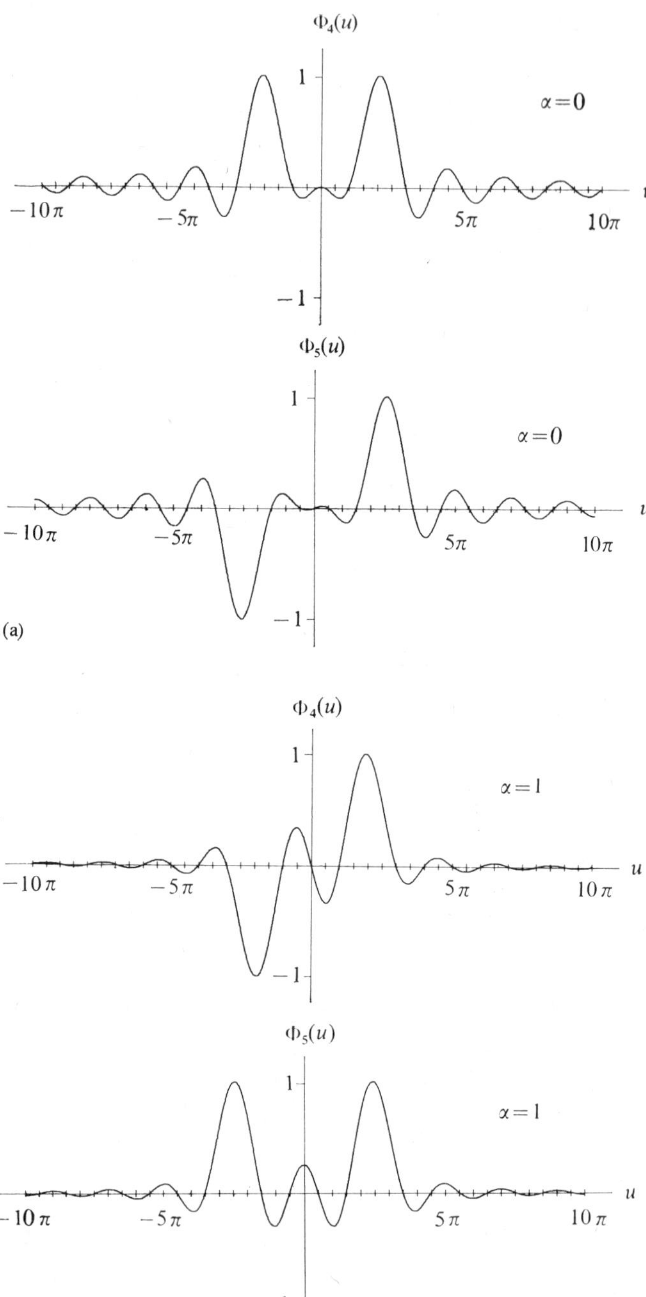

FIG. 6.4. Pattern sampling functions $\Phi_{2n}(u)$ and $\Phi_{2n+1}(u)$ for $n = 2$, for (a) $\alpha = 0$ and (b) $\alpha = 1$.

The coefficients $\{a_p\}$ in expansions (6.16) and (6.17) will now be chosen in such a way as to make the even and odd parts of the pattern space factor $F(u)$ coincide exactly with the even and odd parts of the desired pattern function $\hat{F}(u)$ at alternate values of the sampling points $\{\frac{1}{2}p\pi\}$. From (6.12) it is evident that the value of a_p for $\alpha = 0$ must be chosen to be

$$a_p = \begin{cases} \hat{F}_e(\frac{1}{2}p\pi), & p \text{ even} \\ \hat{F}_o(\frac{1}{2}p\pi), & p \text{ odd} \end{cases} \tag{6.25}$$

and from (6.15) it is evident that the value of a_p for $\alpha = 1$ must be chosen to be

$$a_p = \begin{cases} \hat{F}_o(\frac{1}{2}p\pi), & p \text{ even} \\ \hat{F}_e(\frac{1}{2}p\pi), & p \text{ odd.} \end{cases} \tag{6.26}$$

When the sample values $\hat{F}_e(\frac{1}{2}p\pi)$ and $\hat{F}_o(\frac{1}{2}p\pi)$ of the even and odd parts of $\hat{F}(u)$ are expressed in terms of the sample values of $\hat{F}(u)$ itself these coefficients become

$$a_p = \begin{cases} \frac{1}{2}\{\hat{F}(\frac{1}{2}p\pi)+(-1)^p\hat{F}(-\frac{1}{2}p\pi)\}, & \alpha = 0 \\ \frac{1}{2}\{\hat{F}(\frac{1}{2}p\pi)-(-1)^p\hat{F}(-\frac{1}{2}p\pi)\}, & \alpha = 1. \end{cases} \tag{6.27}$$

The difference between the new source (6.16) and Woodward's is now fully evident. The even part (6.10*a*) for the case $\alpha = 0$ and the odd part (6.13*b*) for the other case $\alpha = 1$ are of precisely the same form as Woodward's even and odd parts (6.9*a*) and (6.9*b*), respectively, when the coefficients $\{a_{2n}\}$ are taken to be zero for $n > N$. Thus, Woodward synthesis is a hybrid case involving only the even coefficients $\{a_{2n}\}$, related to the coefficients $\{b_n\}$ in (6.2) by

$$a_{2n} = \begin{cases} \frac{1}{2}(b_n+b_{-n}), & \alpha = 0 \\ \frac{1}{2}(b_n-b_{-n}), & \alpha = 1. \end{cases} \tag{6.28}$$

It cannot be used to represent completely general sources for either $\alpha = 0$ or 1 but is limited exclusively to even functions for $\alpha = 0$ and to odd functions for $\alpha = 1$. This is a consequence of the fact that it utilizes only one half of the pattern samples required to describe arbitrary aperture-limited functions of either order, namely, those at even multiples of $\frac{1}{2}\pi$, while ignoring completely the other half at odd multiples of $\frac{1}{2}\pi$.

An interesting example of a practical case in which Woodward sampling could not be used was the design by Bayliss† and by Minkovich‡ of a nearly optimum monopulse difference (odd) pattern for $\alpha = 0$ that is analogous to the Taylor sum (even) pattern. Bayliss and Minkovich independently designed the difference pattern to be an odd aperture-limited function of

† E. T. Bayliss. Design of monopulse antenna difference patterns with low sidelobes. *Bell System Tech. J.*, **47**, pp. 623–50, May-June 1968.

‡ B. M. Minkovich. Optimal differential radiation patterns and the synthesis of linear apertures. *Radio Engng electron. Phys.*, **13**, pp. 1046–53, July 1968.

order $\alpha = 0$ with zeros positioned to give maximum slope at the origin for an arbitrarily prescribed sidelobe ratio. Both of them represented the pattern space factor by an expansion in a series of odd sampling functions that are identical with those in (6.21), and the source by an expansion in a series of odd aperture basis functions that are identical with those in (6.18).

6.4. A general theory[14]

There are only two possible values for the edge exponent α in the case of periodic sampling, zero or one. This raises the question as to whether a completely general theory of sampling synthesis might not be possible for arbitrary values of α and for almost arbitrary location of the sampling points. It appears that the answer may well be affirmative. A general theory of sampling synthesis will be developed here that is definitely known to be valid over a limited range of values of α and of location of the sampling points. And there is some indication that it continues to be valid outside of that range. When the sampling points are spaced periodically, as they were in section 6.3, the general sampling functions will be seen to reduce to one or the other of the two complete sets of sampling functions $\{\Phi_n(u)\}$ obtained there for $\alpha = 0$ and 1. Hence the general theory is clearly valid in both of these special cases, even though the value $\alpha = 1$ results in an asymptotic location of the sampling points that is definitely outside of the known range of validity of the general theory. This suggests that the general theory may indeed be valid for all α and for almost any location of the sampling points.

When the sampling points are not spaced periodically the general sampling functions do not form an orthogonal set, in general. It will be seen, however, that they always reduce to one of a pair of biorthogonal sets and that their aperture basis functions do likewise. This biorthogonality property will be used in section 6.5 to derive an interesting class of aperture basis functions for which the expansion coefficients can be determined by means of semi-periodic sampling. These are the functions that will be required in Chapter 7 to generalize the Taylor distribution from Taylor's original case of $\alpha = 0$ to completely arbitrary values of α.

(i) *Formulation of the general theory*

The general theory of sampling synthesis as formulated herein is a direct adaptation of certain mathematical results obtained by Paley and Wiener† on the theory of nonharmonic Fourier series. The principal difference between the formulation here and that of Paley and Wiener lies in an additional requirement that is imposed by the aperture edge behaviour upon the asymptotic location of the zeros of the pattern space factor. This requirement is different for the even part of the pattern space factor than for the odd part,

† R. E. A. C. Paley and N. Wiener, *loc. cit.*, pp. 114–115.

hence the even and odd parts must be treated separately. Only the even part will be treated here. A similar analysis could be developed for the odd part.

From the analytic nature of entire functions $F(u)$ of the form (5.9) it is known that they can be described entirely by the location of their zeros. Their even and odd parts each have an infinite number of zeros that occur in symmetrical pairs in the complex u-plane, the nth pair $\pm u_n$ of which must tend asymptotically toward the points†

$$\pm u_n \sim \begin{cases} \pm\{n+\tfrac{1}{2}\alpha\}\pi & \text{for even part,} \\ \pm\{n+\tfrac{1}{2}(\alpha+1)\}\pi & \text{for odd part,} \end{cases} \tag{6.29}$$

on the real axis as n approaches infinity. This property of $F(u)$, which was first established by Taylor‡ for the case of even functions, is based upon the fact that any function of the form (5.9) must behave asymptotically with increasing absolute values of u as §

$$F(u) \sim 2^{\alpha+1}\Gamma(\alpha+1)\frac{f_e(1)\cos\{u-\tfrac{1}{2}(\alpha+1)\pi\}+if_o(1)\sin\{u-\tfrac{1}{2}(\alpha+1)\pi\}}{u^{\alpha+1}}\times$$

$$\times\left\{1+O\left(\frac{1}{u}\right)\right\}, \tag{6.30}$$

for Re $u > 0$, where $f_e(1)$ and $f_o(1)$ are the even and odd parts of the aperture space factor $f(t)$ evaluated at the endpoint $t = 1$.

Only even sampling functions are required to represent even pattern space factors. The zeros of the sampling functions will again be restricted to the real axis of the complex u-plane. They will always appear as symmetrical pairs located at all but one of an infinite number of pairs of real sampling points $\{\pm\lambda_n\}$ for $n = 0, 1, 2, 3, \ldots$ (this notation for location of the sampling points differs from that of Paley and Wiener by a factor of π because the aperture interval is taken here to be $(-1, 1)$ instead of $(-\pi, \pi)$). It will be assumed that the upper sampling points $\{+\lambda_n\}$ are distinct and ordered such that

$$\lambda_0 < \lambda_1 < \lambda_2 < \lambda_3 < \ldots, \tag{6.31}$$

with only λ_0 permitted to become zero, and that they tend toward the asymptotic location (6.29) required of the zeros of all even aperture-limited functions of order α,

$$\lambda_n \sim (n+\tfrac{1}{2}\alpha)\pi. \tag{6.32}$$

Except for these restrictions they can be considered to be an arbitrary infinite set of real numbers, as indicated in Fig. 6.5.

† The single zero that appears at the origin of all odd functions will be denoted by $n = 0$.

‡ T. T. Taylor (1955), *loc. cit.*, Theorem IV.

§ By an argument similar to that of Taylor's Theorem III.

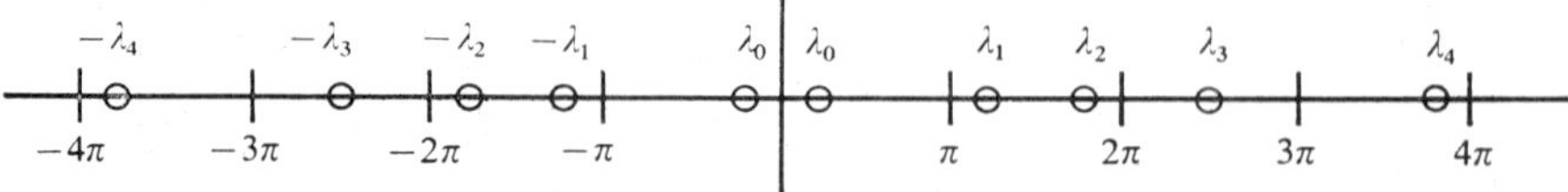

FIG. 6.5. Location of arbitrary pattern-sampling points $\{\pm\lambda_n\}$ along the real axis of the complex u-plane for general sampling synthesis.

An even entire function $G(u)$ whose zeros all lie at the sampling points $\{\pm\lambda_n\}$ can be represented as an infinite product,

$$G(u) = (u^2-\lambda_0^2)\prod_{m=1}^{\infty}\left(1-\frac{u^2}{\lambda_m^{\,2}}\right), \tag{6.33}$$

from which one can construct the *general sampling functions*,

$$\frac{2\lambda_n}{G'(\lambda_n)}\frac{G(u)}{u^2-\lambda_n^{\,2}} \qquad n = 0, 1, 2, 3, \ldots . \tag{6.34}$$

The proportionality constant $2\lambda_n/G'(\lambda_n)$ is included here in order to normalize the sampling functions to unity at the sampling points,

$$\left.\frac{2\lambda_n}{G'(\lambda_n)}\frac{G(u)}{u^2-\lambda_n^{\,2}}\right|_{u=\pm\lambda_p} = \begin{cases}1, & p = n\\ 0, & p \neq n.\end{cases} \tag{6.35}$$

For $\lambda_0 = 0$ the proportionality constant is indeterminate, but from L'Hospital's rule it is seen to be $2/G''(0)$.

Since the zeros of these sampling functions have the asymptotic location required of all aperture-limited functions of order α it is expected that the functions themselves will be the finite Fourier transform of some set of aperture basis functions $\{h_n(t)\}$ whose edge exponent is α,

$$\int_{-1}^{1} h_n(t)\cos ut\,dt = \frac{2\lambda_n}{G'(\lambda_n)}\frac{G(u)}{u^2-\lambda_n^{\,2}}. \tag{6.36}$$

Taking the inverse Fourier transform one obtains the aperture basis functions themselves,

$$h_n(t) = \frac{1}{2\pi}\int_{-\infty}^{\infty}\frac{2\lambda_n}{G'(\lambda_n)}\frac{G(u)}{u^2-\lambda_n^{\,2}}\cos ut\,du, \tag{6.37}$$

which are guaranteed to vanish outside of $(-1, 1)$ and to behave at the edges as the power α of distance from the edge.

Evaluating (6.36) at the sampling point λ_p,

$$\int_{-1}^{1} h_n(t)\cos\lambda_p t\,dt = \left.\frac{2\lambda_n}{G'(\lambda_n)}\frac{G(u)}{u^2-\lambda_n^{\,2}}\right|_{u=\lambda_p} = \delta_{np}, \tag{6.38}$$

it is concluded that *the set of aperture basis functions* $\{h_n(t)\}$ *is biorthonormal to the set of periodic functions* $\{\cos \lambda_n t\}$ *on* $(-1, 1)$.

If the set $\{h_n(t)\}$ is complete with respect to all square-integrable even functions on $(-1, 1)$ then the aperture distribution $A(t)$ can be represented everywhere, including the edges $t = \pm 1$, by

$$A(t) = \sum_{n=0}^{\infty} a_n h_n(t). \tag{6.39}$$

Hence the pattern space factor (6.1) can be represented uniformly on $(-\infty, \infty)$ as an expansion in the general sampling functions (6.34) by transforming the series (6.39) term by term,

$$F(u) = \sum_{n=0}^{\infty} a_n \int_{-1}^{1} h_n(t)\cos ut\, dt = \sum_{n=0}^{\infty} a_n \frac{2\lambda_n}{G'(\lambda_n)} \frac{G(u)}{u^2 - \lambda_n^2}, \tag{6.40}$$

as will be seen shortly.

The expansion coefficients $\{a_n\}$ can be obtained from the aperture distribution $A(t)$ by multiplying both sides of (6.39) by $\cos \lambda_p t$ and integrating over $(-1, 1)$ and then using the biorthonormality relation (6.38),

$$a_n = \int_{-1}^{1} A(t)\cos \lambda_n t\, dt. \tag{6.41}$$

Alternatively, they can be obtained from the pattern space factor $F(u)$ by making use of the sampling property (6.35) of the sampling functions in expansion (6.40),

$$a_n = F(\lambda_n). \tag{6.42}$$

From the Fourier transform relationship (6.1) between the even aperture and pattern functions at $u = \lambda_n$ it is seen that the two values (6.41) and (6.42) are identical.

The above expansions (6.39) for $A(t)$ and (6.40) for $F(u)$ were predicated on the assumption that the set of functions $\{h_n(t)\}$ is complete. It is here that the question arises on the validity of the general sampling method for arbitrary values of α and for arbitrary location of the sampling points. Paley and Wiener showed only that the method is valid whenever λ_0 is zero (a condition that is always satisfied in the case of odd functions but that need not be satisfied for the case of even functions treated here) and when all of the other sampling points lie within the range

$$|\lambda_n - n\pi| < D\pi, \tag{6.43}$$

where $D = 1/\pi^2$. The range was extended somewhat when Levinson†

† N. Levinson. *Gap and density theorems*, Am. Math. Soc. Colloq. Pubs., vol 26, New York, 1940; p. 48.

showed that it was still valid for $D = \frac{1}{4}$. This appears to be the largest range for which completeness has actually been proved. The largest range of α that it would admit is then $-\frac{1}{2} < \alpha < \frac{1}{2}$. It is known that the general sampling theory continues to be valid for $\lambda_0 = \frac{1}{2}\pi$ and $\alpha = 1$, however, from the special case of periodic sampling that was developed in section 6.3. This indicates that the Paley-Wiener and the Levinson conditions may be too restrictive for the problem of interest here. Thus, it is not unreasonable to expect that the general theory developed here will continue to hold for all values of $\alpha > -1$ and all sampling points that satisfy (6.31) and (6.32).

(ii) *Biorthogonality and completeness*

Two important properties of the general sampling functions (6.34) will be established here. One is the fact that they always form one of a pair of biorthogonal sets on $(-\infty, \infty)$. The other is the fact that they are complete with respect to all even aperture-limited functions.

The biorthogonality property is possessed by any set of functions $\{X_n(u)\}$ that is the finite Fourier transform of some other set of functions $\{\chi_n(t)\}$ known to be one of a pair of biorthogonal sets. This is easily demonstrated as follows. Suppose that $\{X_n(u)\}$ and $\{\Psi_n(u)\}$ are the finite Fourier transforms of $\{\chi_n(t)\}$ and $\{\psi_n(t)\}$, respectively,

$$X_n(u) = \int_{-1}^{1} \chi_n(t) e^{iut}\, dt, \tag{6.44a}$$

$$\Psi_n(u) = \int_{-1}^{1} \psi_n(t) e^{iut}\, dt. \tag{6.44b}$$

If $\{\chi_n(t)\}$ and $\{\psi_n(t)\}$ are biorthogonal on $(-1, 1)$,

$$\int_{-1}^{1} \chi_n(t) \psi_p{}^*(t)\, dt = \delta_{np}, \tag{6.45}$$

then $\{X_n(u)\}$ and $\{\Psi_n(u)\}$ are biorthogonal on $(-\infty, \infty)$, and conversely,

$$\int_{-\infty}^{\infty} X_n(u) \Psi_p{}^*(u)\, du = \int_{-1}^{1} \int_{-1}^{1} \chi_n(t) \psi_p{}^*(t') \left\{ \int_{-\infty}^{\infty} e^{iu(t-t')}\, du \right\} dt\, dt'$$

$$= 2\pi \int_{-1}^{1} \chi_n(t) \psi_p{}^*(t)\, dt = 2\pi\, \delta_{np}. \tag{6.46}$$

In the case of the functions appearing in the general theory of sampling the biorthogonal functions of t are

$$\chi_n(t) = h_n(t), \tag{6.47a}$$

$$\psi_n(t) = \cos \lambda_n t, \tag{6.47b}$$

from which the corresponding biorthogonal functions of u are seen to be

$$X_n(u) = \frac{2\lambda_n}{G'(\lambda_n)} \frac{G(u)}{u^2-\lambda_n^2}, \tag{6.48a}$$

$$\Psi_n(u) = \frac{\sin(u-\lambda_n)}{u-\lambda_n} + \frac{\sin(u+\lambda_n)}{u+\lambda_n}. \tag{6.48b}$$

Thus, even though the sampling functions (6.48*a*) may not form an orthogonal set they will always be at least biorthogonal to the functions (6.48*b*).

The proof of completeness, on the other hand, is wholly dependent upon the assumption that the set of aperture basis functions $\{\chi_n(t)\}$ is complete. Completeness of the sampling functions will thus be established by first forming the absolute value of the difference between any given aperture-limited even function and its $N+1$-term approximation as a series of aperture-limited sampling functions $\{X_n(u)\}$,

$$\left|F(u) - \sum_{n=0}^{N} a_n X_n(u)\right| = \left|\int_{-1}^{1} \{A(t) - \sum_{n=0}^{N} a_n \chi_n(t)\}\, e^{iut}\, dt\right|$$

$$\leqslant \int_{-1}^{1} \left|A(t) - \sum_{n=0}^{N} a_n \chi_n(t)\right| dt, \tag{6.49}$$

in which the coefficients $\{a_n\}$ are chosen to have the value (6.41) obtained from the biorthogonality property of the $\{\chi_n(t)\}$ functions. If the set $\{\chi_n(t)\}$ is complete with respect to all even square-integrable functions on $(-1, 1)$ then the latter integral must approach zero as N approaches infinity. The absolute value of the difference between the aperture-limited function and its $N+1$-term approximation must then approach zero uniformly on $-\infty < u < \infty$. Thus it is concluded that the set of even aperture-limited sampling functions $\{X_n(u)\}$ is complete with respect to all even aperture-limited functions.

(iii) *Self-orthogonal sampling functions*

The fact that the sampling functions $\{\Phi_p(u)\}$ obtained for the two special cases of periodic sampling were orthogonal to each other, and not merely to some other functions of a biorthogonal set, raises the question as to whether there might be other sets of sampling functions that are also self-orthogonal. It will now be shown that there are an infinite number of other sets of self-orthogonal sampling functions, but that they are all members of a rather special, and somewhat strange, class in which no two pairs of sampling points can have the same spacing.

A necessary and sufficient condition for self-orthogonality of a set of sampling functions is that their biorthogonal functions be proportional to them. This follows from the completeness property established above for the set of sampling functions $\{X_p(u)\}$, since any function $\Psi_n(u)$ that is bi-orthogonal to all but the single function $X_n(u)$ of a complete set of self-orthogonal functions $\{X_p(u)\}$ must be proportional to that function $X_n(u)$. And if the two are proportional their zeros must all coincide. Since the qth sampling point λ_q denotes a zero of $X_n(u)$ for all $q \neq n$ then it must also denote a zero of $\Psi_n(u)$; i.e.

$$\Psi_n(\lambda_q) = 0, \qquad q \neq n. \tag{6.50}$$

From the definition (6.48*b*) for $\Psi_n(u)$ this requires that

$$\frac{2}{\lambda_q^{\,2} - \lambda_n^{\,2}}(\lambda_q \sin \lambda_q \cos \lambda_n - \lambda_n \sin \lambda_n \cos \lambda_q) = 0, \qquad q \neq n, \tag{6.51}$$

or that

$$\lambda_q \tan \lambda_q = \lambda_n \tan \lambda_n. \tag{6.52}$$

For this to hold for all n and q it follows that both sides must be constant. Hence the sampling functions $\{X_p(u)\}$ will be self-orthogonal if, and only if, the sampling points are the roots of the transcendental equation

$$\lambda_q \tan \lambda_q = C, \tag{6.53}$$

where C is some arbitrary real constant.

At one extreme, $C = 0$, the roots of (6.53) are seen to be

$$\lambda_q = 0, \pi, 2\pi, 3\pi, \dots, \tag{6.54}$$

which are simply the set of periodic sampling points that defined the even sampling functions $\{\Phi_{2n}(u)\}$ for $\alpha = 0$. At the other extreme, $C = \infty$, the roots are

$$\lambda_q = \tfrac{1}{2}\pi, \tfrac{3}{2}\pi, \tfrac{5}{2}\pi, \dots, \tag{6.55}$$

which are the set of periodic sampling points that defined the even sampling functions $\{\Phi_{2n+1}(u)\}$ for $\alpha = 1$. For any other value of C lying between these two extremes there must also be an infinite set of roots $\{\lambda_q\}$ that represent the sampling points of some other set of even sampling functions that are orthogonal to each other. The spacing between successive roots will always decrease monotonically with increasing values of q, and the qth root will always tend toward the value $q\pi$ as q approaches infinity. Hence it is evident from (6.29) that for any finite value of C these strange self-orthogonal functions will always be sampling functions of order $\alpha = 0$ that sample nonperiodically. Over any finite interval, however, they can be made arbitrarily close to the sampling functions $\{\Phi_{2n+1}(u)\}$ of order $\alpha = 1$ that sample periodically, simply by choosing the value of C to be sufficiently large.

The aperture basis functions that produce these self-orthogonal sampling functions must themselves be self-orthogonal. This follows from (6.46). It was true for the functions $\{\phi_{2n}(t)\}$ for $\alpha = 0$ that were obtained when C

is zero and for the functions $\{\phi_{2n+1}(t)\}$ for $\alpha = 1$ that were obtained when C is infinite, and it must be equally true for the other sets of aperture basis functions that are obtained for other values of C.

6.5. Semi-periodic sampling synthesis[14]

The functions arising from the general theory will now be investigated in more detail for the special case of semi-periodic sampling. By semi-periodic sampling will be meant that the spacing between the upper sampling points $\{+\lambda_n\}$ (or between the lower sampling points $\{-\lambda_n\}$) is exactly π.

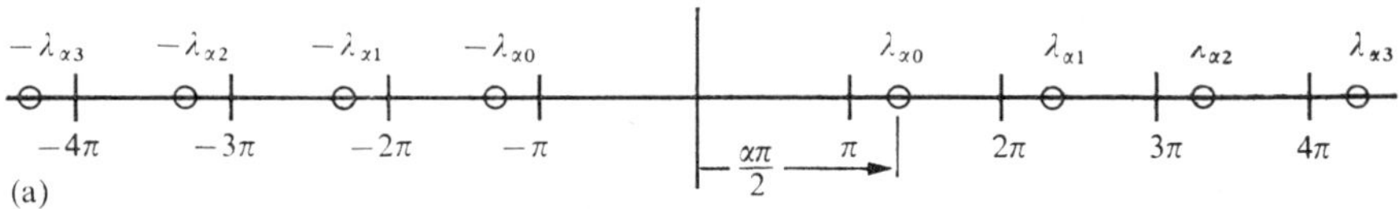

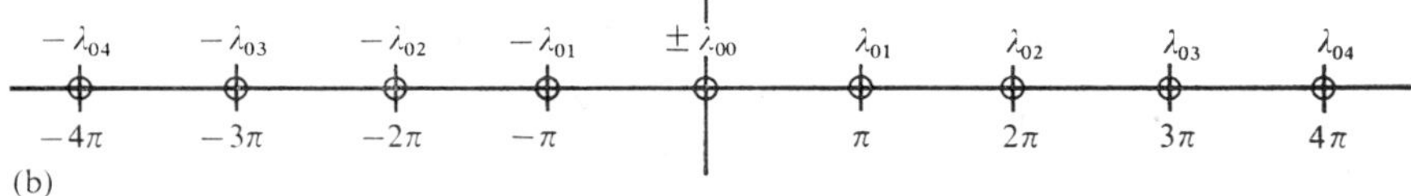

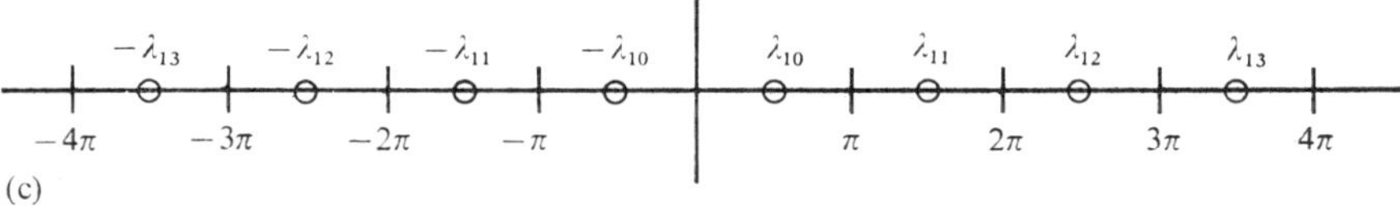

FIG. 6.6. Semi-periodic sampling points $\lambda_{\alpha n} = (n+\frac{1}{2}\alpha)\pi$ for (a) arbitrary $\alpha > -1$, (b) $\alpha = 0$, and (c) $\alpha = 1$.

If, in addition, the spacing between the innermost upper and lower sampling points $\pm\lambda_0$ is exactly 0 or π then semi-periodic sampling reduces to the even more special case of periodic sampling developed in section (6.3).

In view of the asymptotic behaviour (6.29) that the zeros of the pattern space factor must have for any given value of α it follows that the location of the upper sampling points for semi-periodic sampling of even pattern space factors must be

$$\lambda_{\alpha n} = (n+\tfrac{1}{2}\alpha)\pi, \qquad n = 0, 1, 2, 3, \ldots, \tag{6.56}$$

for any $\alpha > -1$. This is illustrated in Fig. 6.6a. The only cases in which the sampling points become truly periodic occur when α is 0 or 1, as in Figs. 6.6b and 6.6c, respectively. These are the two cases that were investigated earlier in section 6.3.

The semi-periodic sampling functions obtained from (6.34) for $\alpha > -1$ are readily found to be

$$\frac{2\lambda_{\alpha n}}{G_\alpha'(\lambda_{\alpha n})}\frac{G_\alpha(u)}{u^2-\lambda_{\alpha n}{}^2} = (-1)^{n+1}\frac{(2n+\alpha)\pi^2\Gamma(n+\alpha)}{n!\,\Gamma(\frac{1}{2}\alpha+(u/\pi))\Gamma(\frac{1}{2}\alpha-(u/\pi))\{u^2-(n+\frac{1}{2}\alpha)^2\pi^2\}}$$
$$= (-1)^n\frac{(2n+\alpha)\pi\Gamma(n+\alpha)}{n!}\frac{\Gamma((u/\pi)-\frac{1}{2}\alpha+1)\sin(u-\frac{1}{2}\alpha\pi)}{\Gamma((u/\pi)+\frac{1}{2}\alpha)\{u^2-(n+\frac{1}{2}\alpha)^2\pi^2\}}, \tag{6.57}$$

by first expressing the infinite product (6.33) in terms of Gamma functions† and then using certain well known properties of the Gamma functions.

Finding the aperture basis functions $\{h_{\alpha n}(t)\}$ required to produce these sampling functions is a bit more difficult. The inverse Fourier transform (6.37) is, from (6.57),

$$h_{\alpha n}(t) = (-1)^n\frac{(n+\frac{1}{2}\alpha)\Gamma(n+\alpha)}{n!}\int_{-\infty}^{\infty}\frac{\cos \pi tz\, dz}{\Gamma(\frac{1}{2}\alpha+z)\Gamma(\frac{1}{2}\alpha-z)\{(\frac{1}{2}\alpha+n)^2-z^2\}}. \tag{6.58}$$

For $n = 0$ the integral can be evaluated rather simply from the known Fourier transform‡

$$\int_0^1 (\cos \tfrac{1}{2}\pi t)^\alpha \cos \pi tz\, dt = \frac{\Gamma(\alpha+1)}{2^\alpha\Gamma(\frac{1}{2}\alpha+1+z)\Gamma(\frac{1}{2}\alpha+1-z)} \tag{6.59}$$

for all $\alpha > -1$, the inverse of which gives

$$\int_{-\infty}^{\infty}\frac{\cos \pi tz\, dz}{\Gamma(\frac{1}{2}\alpha+1+z)\Gamma(\frac{1}{2}\alpha+1-z)} = \begin{cases}\dfrac{1}{\Gamma(\alpha+1)}(2\cos\frac{1}{2}\pi t)^\alpha, & 0 < t < 1\\ 0, & t > 1.\end{cases} \tag{6.60}$$

For all other values of n it can be evaluated from (6.60) by making use of the recursion property of the Gamma functions and then reducing it to a sum of second derivatives of integrals of the form (6.60). When that procedure was used to obtain the first few functions it was observed that they all have a common functional form, namely, the factor $(2\cos\frac{1}{2}\pi t)^\alpha$ times an $n+1$-term series of $\cos p\pi t$ functions. By expressing $h_{\alpha n}(t)$ as such a series with unknown coefficients and then imposing the biorthogonality condition (6.38) the coefficients were found to be simply $(-1)^{n-p}\frac{1}{2}\epsilon_p$ times a generalization of the binomial coefficients. Thus, the aperture basis functions (6.58) required to produce the semi-periodic sampling functions (6.57) for

† I. S. Gradshteyn and I. M. Ryzhik. *Table of integrals, series, and products*, Academic Press, New York, 1965; eqn. (8.325.1).

‡ Gradshteyn and Ryzhik, *loc. cit.* eqn (3.631.9).

any $\alpha > -1$ can be expressed in closed form as

$$h_{\alpha n}(t) = (2\cos\tfrac{1}{2}\pi t)^{\alpha}\sum_{p=0}^{n}(-1)^{n-p}\tfrac{1}{2}\epsilon_p\frac{\Gamma(\alpha+n-p)}{(n-p)!\Gamma(\alpha)}\cos p\pi t \tag{6.61}$$

for $0 < t < 1$, and zero for $t > 1$.

In the two special cases of true periodic sampling, namely, $\alpha = 0$ and 1, the semi-periodic sampling functions (6.57) reduce to

$$\frac{2\lambda_{\alpha n}}{G_{\alpha}'(\lambda_{\alpha n})}\frac{G_{\alpha}(u)}{u^2-\lambda_{\alpha n}{}^2} = \begin{cases}(-1)^n\tfrac{1}{2}\epsilon_n\dfrac{2u\sin u}{u^2-\{n\pi\}^2}, & \alpha = 0\\[2ex] (-1)^{n+1}\dfrac{(2n+1)\pi\cos u}{u^2-\{(n+\frac{1}{2})\pi\}^2}, & \alpha = 1,\end{cases} \tag{6.62}$$

and the aperture basis functions (6.61) reduce to

$$h_{\alpha n}(t) = \begin{cases}\tfrac{1}{2}\epsilon_n\cos n\pi t, & \alpha = 0\\ \cos(n+\frac{1}{2})\pi t, & \alpha = 1.\end{cases} \tag{6.63}$$

These are identical to the even sampling functions $\{\Phi_{2n}(u)\}$ in (6.21) for $\alpha = 0$ and $\{\Phi_{2n+1}(u)\}$ in (6.22) for $\alpha = 1$, and to the even aperture basis functions $\{\phi_{2n}(t)\}$ in (6.18) for $\alpha = 0$ and $\{\phi_{2n+1}(t)\}$ in (6.19) for $\alpha = 1$.

Another important special case is the one for which the edge exponent has the value $\alpha = 2$ required for all sources with physically realizable edge behaviour in the H-plane. The sampling functions (6.57) and the aperture basis functions (6.61) in that case become

$$\frac{2\lambda_{2n}}{G_2'(\lambda_{2n})}\frac{G_2(u)}{u^2-\lambda_{2n}{}^2} = (-1)^{n+1}\frac{2(n+1)^2\pi^2\sin u}{u[u^2-\{(n+1)\pi\}^2]}, \tag{6.64}$$

$$h_{2n}(t) = \cos(n+1)\pi t+(-1)^n. \tag{6.65}$$

In contrast to the functions for the two periodic cases $\alpha = 0$ and 1 these functions for $\alpha = 2$ do not form orthogonal sets. This is easily demonstrated by direct calculation. For example, for the aperture basis functions,

$$\int_{-1}^{1} h_{2n}(t)h_{2p}(t)\,dt = \delta_{np}+2(-1)^{n-p}. \tag{6.66}$$

But they still must form one of a pair of biorthogonal sets.

CHAPTER 7

MAXIMUM RESOLUTION[10]

BY the resolution of an antenna is meant its ability to discriminate against signals arriving from directions other than some specified direction. Resolution is frequently described in terms of beamwidth, an antenna with a narrow beam being said to have more resolution than one whose beam is wider. Such a description is valid as far as it goes but it doesn't face up to the problem of discriminating against signals received on sidelobes. An antenna whose main beam is made narrower at the expense of an unacceptable increase in sidelobe level can hardly be said to have greater resolution. In the case of most optical apertures, for which the Rayleigh criterion was formulated originally, there is little or no control over sidelobe level because very little control is available over the aperture distribution. About all that can be done at optical frequencies is to make the phase of the aperture distribution as nearly uniform as possible. At radio frequencies, however, the aperture distribution can be controlled much more precisely, and not just in phase but in magnitude as well. With this additional freedom it becomes quite practical to exercise direct control over the sidelobe level. Sidelobes can never be wholly eliminated, of course, because of analyticity of the pattern space factor, but they can always be forced to lie below some acceptable level relative to the strength of the main beam. Even so, a reduction in sidelobe level is not entirely free of cost, as it can be obtained only at the expense of an increase in beamwidth or in Q or in both together. Thus, by an optimum solution to the maximum resolution problem must be meant one that results in the minimum beamwidth possible for arbitrarily specified values of the sidelobe ratio (the ratio of the strength of the main beam to that of the largest sidelobe) and of Q.

No optimum solution to the maximum resolution problem is known as yet. But a remarkable near-optimum solution was constructed by Taylor,† based on a philosophy and a synthesis technique that was totally new at the time. It appears to have been the first time that source synthesis was recognized as being reducible to a problem of positioning the zeros of an entire transcendental function of exponential type. This philosophy of describing the pattern space factor in terms of the location of its zeros has since become an integral part of modern antenna theory, going well beyond the maximum resolution problem itself. It has already been used here to develop the general theory of sampling synthesis described in section 6.4. It will now

† T. T. Taylor (1955), *loc. cit.*

be used again to extend Taylor's near-optimum solution of the maximum resolution problem from his special case of $\alpha = 0$ to the general case in which α can be any real number greater than -1.

7.1. The general Taylor pattern

Taylor based the construction of his near-optimum pattern space factor on three points. The first was his proof that any even aperture-limited function of order α is an entire transcendental function of u whose nth pair of zeros tend to the real positions $\pm(n+\frac{1}{2}\alpha)\pi$ as n tends to infinity. The second was his contention that the precision with which an aperture-limited pattern space factor can be controlled over the visible region will become greatest as the value of α approaches its lower limit -1, on the grounds that it would then provide the greatest number of 'central zeros'. And the third was his assumption that the van der Maas function† $\cos(u^2-\pi^2A^2)^{\frac{1}{2}}$ is the 'ideal' pattern space factor, because of the fact that it is an even entire transcendental function of u whose nth pair of zeros tend to the positions $\pm(n-\frac{1}{2})\pi$ (corresponding to the lower limit $\alpha = -1$) and whose beamwidth between first nulls is the smallest that is possible for a prescribed sidelobe ratio $\cosh \pi A$ over the entire real axis. On the further assumption that all non-negative values of α are physically realizable he concluded that the smallest non-negative value, $\alpha = 0$, is the best choice. He then constructed an even aperture-limited function of order $\alpha = 0$ whose first $\bar{n}$ pair of zeros was positioned in such a way as to approximate that of the ideal van der Maas pattern. Thus, the Taylor pattern provides an aperture-limited approximation of order $\alpha = 0$ to this ideal pattern over the central region $n \leqslant \bar{n}$, the accuracy of which can be made arbitrarily high over any given interval by making the parameter $\bar{n}$ sufficiently large.

That the van der Maas function does, in fact, have the smallest possible beamwidth for a given sidelobe ratio over the entire real axis follows from the fact that it is the asymptotic form of a Chebyshev polynomial of Nth degree as N approaches infinity,

$$T_N\left(\cosh\frac{\pi A}{N}-\frac{u^2}{2N^2}\right) = \cos\left\{N \operatorname{arc}\cos\left(\cosh\frac{\pi A}{N}-\frac{u^2}{2N^2}\right)\right\}$$
$$\sim \cos(u^2-\pi^2A^2)^{\frac{1}{2}}. \tag{7.1}$$

For any integer N it can be established‡ that this polynomial is the polynomial of Nth degree that results in the optimum relationship between beamwidth and sidelobe ratio within the real interval

$$|u| \leqslant N[2\{\cosh(\pi A/N)+1\}]^{\frac{1}{2}}.$$

† G. J. van der Maas. A simplified calculation for Dolph–Tchebycheff arrays. *J. appl. Phys.* **25**, pp. 121–4, January 1954.

‡ R. Courant and D. Hilbert, *loc. cit.*, p. 89.

When N becomes infinite this interval covers all of the real axis and the polynomial becomes an entire transcendental function of u whose nth pair of zeros $\pm\{A^2+(n-\frac{1}{2})^2\}^{\frac{1}{2}}\pi$ tend to the positions $\pm(n-\frac{1}{2})\pi$ as n tends to infinity, coinciding precisely with the asymptotic positions (6.29) of the zeros of an even aperture-limited function of order $\alpha = -1$. This is illustrated graphically in Fig. 7.1.

In spite of the square root the van der Maas function is an entire function of u. This can be seen from the fact that its power series expansion

$$\cos(u^2-\pi^2A^2)^{\frac{1}{2}} = 1-\frac{u^2-\pi^2A^2}{2!}+\frac{(u^2-\pi^2A^2)^2}{4!}-\frac{(u^2-\pi^2A^2)^3}{6!}+\dots \tag{7.2}$$

is single-valued and converges for all finite values of u. Furthermore its zeros $\pm u_n$, where

$$u_n = \{A^2+(n-\tfrac{1}{2})^2\}^{\frac{1}{2}}\pi, \tag{7.3}$$

are all simple. Consequently it can be expressed as an infinite product in terms of those zeros,†

$$\cos(u^2-\pi^2A^2)^{\frac{1}{2}} = \cosh \pi A \prod_{n=1}^{\infty}\left[1-\frac{u^2}{\{A^2+(n-\frac{1}{2})^2\}\pi^2}\right]. \tag{7.4}$$

The Taylor near-optimum pattern space factor for arbitrary $\alpha > -1$ can now be constructed by shifting outward the zeros of (7.4) for all $n \geqslant \bar{n}$ to the new positions $u_n = (n+\frac{1}{2}\alpha)\pi$, in order to produce the required asymptotic behaviour (6.29), and then dilating linearly the distance from the origin to all of the remaining zeros, in order to obtain nearly equal sidelobes in the central region $u < u_{\bar{n}}$. The zeros in the dilated central region then become $\pm u_n$, where

$$u_n = \sigma_\alpha\{A^2+(n-\tfrac{1}{2})^2\}^{\frac{1}{2}}\pi, \tag{7.5}$$

in which the dilation constant σ_α required to shift the $\bar{n}$th zero pair linearly from its old position $\pm\{A^2+(\bar{n}-\frac{1}{2})^2\}^{\frac{1}{2}}\pi$ to the new position $\pm(\bar{n}+\frac{1}{2}\alpha)\pi$ is

$$\sigma_\alpha = \frac{\bar{n}+\frac{1}{2}\alpha}{\{A^2+(\bar{n}-\frac{1}{2})^2\}^{\frac{1}{2}}}. \tag{7.6}$$

Thus, the *general Taylor pattern* function $T_\alpha(u, A, \bar{n})$ for any $\alpha > -1$ can be written‡

$$T_\alpha(u, A, \bar{n}) = \prod_{n=1}^{\bar{n}-1}\left[1-\frac{u^2}{\sigma_\alpha^{\,2}\{A^2+(n-\frac{1}{2})^2\}\pi^2}\right]\cdot\prod_{n=\bar{n}}^{\infty}\left[1-\frac{u^2}{(n+\frac{1}{2}\alpha)^2\pi^2}\right]. \tag{7.7}$$

† E. T. Whittaker and G. N. Watson, *loc. cit.*, p. 137.

‡ The general Taylor pattern function is defined here to have unit value in the broadside direction $u = 0$. It differs from Taylor's definition by a constant factor $\cosh \pi A$.

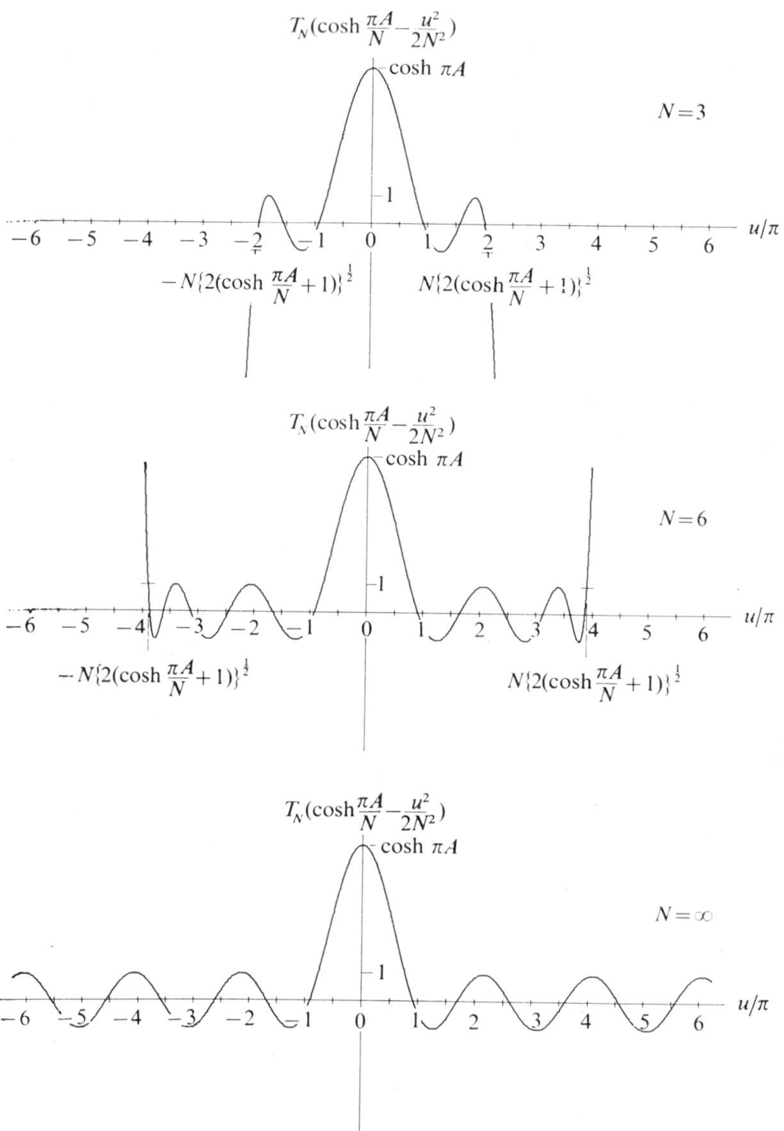

FIG. 7.1. The Chebyshev function $T_N(\cosh[\pi A/N]-u^2/2N^2)$ for a 15dB sidelobe ratio ($A^2 = 0{\cdot}58950$) with $N = 3$, 6, and ∞. For $N = \infty$ the Chebyshev function reduces simply to $\cos(u^2-\pi^2A^2)^{\frac{1}{2}}$. Note that the main beam and the position of the first zero remain essentially unchanged for $N \geqslant 3$.

By expressing the infinite product in terms of Gamma functions, as in (6.57), the Taylor pattern can be expressed more simply as just a ratio of finite products,

$$T_\alpha(u, A, \bar{n}) = \frac{\Gamma^2(\frac{1}{2}\alpha+1)\Gamma((u/\pi)-\frac{1}{2}\alpha+1)}{\Gamma((u/\pi)+\frac{1}{2}\alpha+1)} \frac{\sin(u-\frac{1}{2}\alpha\pi)}{u-\frac{1}{2}\alpha\pi} \times$$

$$\times \frac{\prod_{n=1}^{\bar{n}-1}\left[1-\dfrac{u^2}{\sigma_\alpha{}^2\{A^2+(n-\frac{1}{2})^2\}\pi^2}\right]}{\prod_{n=1}^{\bar{n}-1}\left[1-\dfrac{u^2}{(u+\frac{1}{2}\alpha)^2\pi^2}\right]}. \tag{7.8}$$

It is an entire function of u for any preassigned value of the edge exponent α, of the sidelobe ratio $\cosh \pi A$, and of the integer $\bar{n}$ that separates the central region $u < u_{\bar{n}}$ of nearly equal sidelobes from the outer region $u > u_{\bar{n}}$ of sidelobes that decay asymptotically as $1/u^{\alpha+1}$. For the special case of $\alpha = 0$ it is proportional to Taylor's original pattern function.

The beamwidth between first nulls is greater than that of the ideal pattern by a factor of exactly σ_α. The factor σ_α approaches unity as $\bar{n}$ approaches infinity, as expected since the zeros of the Taylor function then approach those of the ideal function. But even for relatively small values of $\bar{n}$ the value of σ_α is not much greater than unity for physically interesting values of α. For example, for $\bar{n} = 10$ and $\alpha = 0$ the beamwidth for a sidelobe ratio of 20 dB($A^2 = 0{\cdot}90777$) will be greater than the ideal by a factor of only $\sigma_0 = 1{\cdot}047$, or by 4·7 per cent.

The beamwidth increases also with α. It will be greater than that for $\alpha = 0$, for example, by the factor

$$\frac{\sigma_\alpha}{\sigma_0} = 1+\frac{\alpha}{2\bar{n}}, \tag{7.9}$$

from (7.6), independently of the value of the sidelobe ratio. Thus, for the case of $\bar{n} = 10$ the beamwidth for $\alpha = 1$ will be greater than that for $\alpha = 0$ by only 5 per cent.

By choosing a sufficiently large value for the parameter $\bar{n}$ it is now evident that the Taylor pattern can be made arbitrarily close to the ideal van der Maas pattern for any given value of α. For that reason it is generally believed that the Taylor relationship between beamwidth and sidelobe ratio is close to being the best that can be achieved. While such a conclusion is probably true it should not go unrecognized that the van der Maas idealization contains an implicit assumption that has no physical basis in fact and that may cause this idealization to fall short of the true ideal. Specifically, there is no physical basis for the assumption that the invisible lobes must remain within the prescribed level of the visible sidelobes, since invisible lobes do not represent sidelobes in the usual sense. They are not even a part of the radiation pattern, their only connection with the radiation pattern being through the mathematical process of analytic continuation. Their physical interpretation is in terms of reactive, not radiative, power. By unnecessarily constraining the invisible lobes to remain within the prescribed sidelobe level the resulting relationship between beamwidth and sidelobe ratio of the van der Maas pattern, and thence of the Taylor pattern, will surely be less than optimum. The only valid physical constraint on the invisible lobes is that which is imposed by the quality factor Q. Since the physical constraint imposed by Q need not be as stringent as the artificial constraint imposed by the sidelobe ratio it is evident that a truly optimum solution to the maximum resolution problem, if one were obtained, could result in a narrower beamwidth for a given sidelobe ratio and value of Q than does Taylor's near-optimum solution.

7.2. The general Taylor distribution

The general Taylor pattern (7.8) was constructed to have the same asymptotic behaviour as the class of all even aperture-limited functions of order α. Hence it follows that (7.8) must be the finite Fourier transform of some aperture distribution with an arbitrarily prescribed value of the edge exponent α. In principle that aperture distribution can be calculated exactly by taking the inverse Fourier transform. The problem now is to find some practical way to do it.

Taylor did it by using Woodward's sampling method described in section 6.1, but that was possible only because the edge exponent for the even part of Woodward's aperture distribution happened to have precisely the value $\alpha = 0$ chosen by Taylor. It would not have been possible for any other value of α, nor would it have been possible if Taylor's pattern function had had an odd part, as in the case, for example, of a Taylor pattern scanned away from the broadside direction. Instead, the appropriate sampling method for arbitrary α is the generalization that was developed for semi-periodic sampling in section 6.5. This follows from the fact that the zeros of the general Taylor pattern for all $n \geqslant \bar{n}$ lie at the semi-periodic sampling points $\pm(n+\frac{1}{2}\alpha)\pi$. Hence, if the general Taylor pattern is expanded in a biorthogonal series of the semi-periodic sampling functions (6.57) all of its expansion coefficients for $n \geqslant \bar{n}$ will be zero, leaving only a finite series that terminates with the $\bar{n}-1$ term. The value of each of the $\bar{n}$ nonzero expansion coefficients will be just the value (7.8) of $T_\alpha(u, A, \bar{n})$ at each of the sampling points $(n+\frac{1}{2}\alpha)\pi$.

Thus, the general Taylor distribution for any real value of $\alpha > -1$ can be represented exactly by the series (6.39) of biorthogonal aperture basis functions (6.61) whose $\bar{n}-1$ nonzero coefficients are the sample values of the Taylor pattern (7.8).

7.3. Illustration for $\alpha = 0$ and 1

The Taylor near-optimum solution of the maximum resolution problem will be illustrated here for two different cases, $\alpha = 0$ and 1, the first of which is the one treated originally by Taylor. Taylor's original case cannot be realized exactly in the E-plane, nor can it even be approximated in the H-plane. For the new case $\alpha = 1$, on the other hand, the near-optimum solution can be realized exactly in the E-plane, and for $\alpha = 0$ and 1 in the E- and H-planes, respectively, it can be approximated as accurately as desired. Hence these appear to be the two cases of primary practical interest.

When α is 0 or 1 the general Taylor pattern (7.8) reduces to

$$T_0(u, A, \bar{n}) = \frac{\sin u}{u} \frac{\prod_{n=1}^{\bar{n}-1}\left[1-\frac{u^2}{\sigma_0^2\{A^2+(n-\frac{1}{2})^2\}\pi^2}\right]}{\prod_{n=1}^{\bar{n}-1}\left[1-\frac{u^2}{n^2\pi^2}\right]}, \tag{7.10}$$

$$T_1(u, A, \bar{n}) = \frac{\cos u}{1-(2u/\pi)^2} \frac{\prod_{n=1}^{\bar{n}-1}\left[1-\frac{u^2}{\sigma_1^2\{A^2+(n-\frac{1}{2})^2\}\pi^2}\right]}{\prod_{n=1}^{\bar{n}-1}\left[1-\frac{u^2}{(n+\frac{1}{2})^2\pi^2}\right]}. \tag{7.11}$$

For $\alpha = 0$ it can be considered to be the function $(\sin u)/u$ with its first $\bar{n}-1$ zeros shifted to coincide with those of the dilated van der Maas function, while for $\alpha = 1$ it can be considered to be the function $(\cos u)/\{1-(2u/\pi)^2\}$ with its first $\bar{n}-1$ zeros shifted likewise.

For $\alpha = 0$ and 1 the semi-periodic sampling points become periodic, resulting in the particularly simple case of periodic sampling that was developed in section 6.3. Hence the aperture distribution and the pattern space factor for either $\alpha = 0$ or 1 can be expressed exactly by

$$A(t) = \sum_{p=0}^{\infty} a_p(\tfrac{1}{2}\epsilon_p)\cos \tfrac{1}{2}p\pi t, \tag{7.12}$$

$$F(u) = \sum_{p=0}^{\infty} a_p \frac{\epsilon_p}{2}\left\{\frac{\sin(u-\frac{1}{2}p\pi)}{u-\frac{1}{2}p\pi}+\frac{\sin(u+\frac{1}{2}p\pi)}{u+\frac{1}{2}p\pi}\right\}, \tag{7.13}$$

from (6.16) and (6.17), respectively, where the expansion coefficients $\{a_p\}$ for $\alpha = 0$ are the sample values of $T_0(u, A, \bar{n})$ at all integral multiples of π,

$$a_p = \begin{cases} T_0(\frac{1}{2}p\pi, A, \bar{n}), & p \text{ even} \\ 0, & p \text{ odd}, \end{cases} \tag{7.14}$$

while those for $\alpha = 1$ are the sample values of $T_1(u, A, \bar{n})$ at all half-integral multiples of π,

$$a_p = \begin{cases} 0, & p \text{ even} \\ T_1(\tfrac{1}{2}p\pi, A, \bar{n}), & p \text{ odd.} \end{cases} \tag{7.15}$$

In both cases the sampling points for $p \geqslant 2\bar{n}$ coincide with zeros of the Taylor pattern, hence both series terminate after $\bar{n}$ terms. Once the $\bar{n}$ nonzero expansion coefficients have been evaluated from (7.10) or (7.11) both the aperture distribution and the Taylor pattern itself can be computed exactly and very simply from (7.12) and (7.13), respectively.

The Taylor pattern and its aperture distribution for $\alpha = 0$ and 1 are shown superposed in Fig. 7.2 for $\bar{n} = 5$ with design sidelobe ratios of 10, 15, 20,

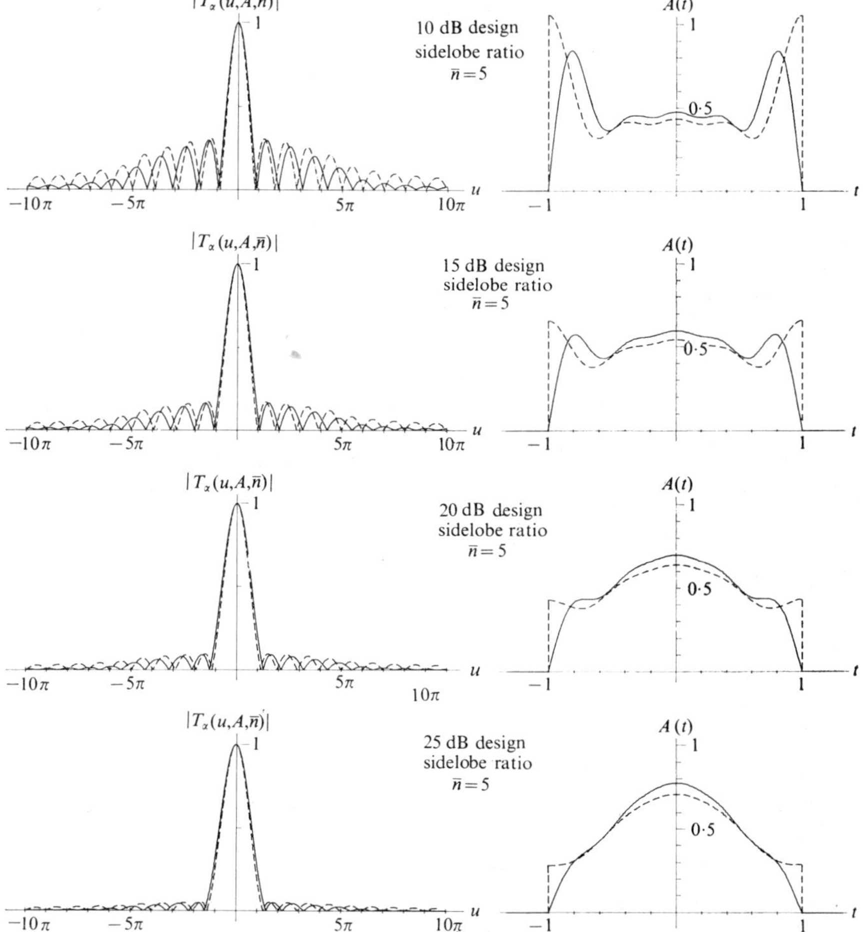

FIG. 7.2. The Taylor pattern space factor $T_\alpha(u, A, \bar{n})$ and its aperture distribution $A(t)$ for $\alpha = 0$ (broken) and $\alpha = 1$ (solid) for $\bar{n} = 5$ with design sidelobe ratios of 10, 15, 20, and 25 dB.

and 25 dB. When the value of the parameter $\bar{n}$ is increased to $\bar{n} = 10$ the pattern and aperture functions for the 20 dB design sidelobe ratio change to those shown in Fig. 7.3. Note that the aperture distributions for $\alpha = 0$ all terminate at the edges in a pedestal while those for $\alpha = 1$ go to zero linearly, as required for those particular values of the edge exponent. And the envelope of the pattern space factor is seen to decay more rapidly with increasing values of u for $\alpha = 1$ than for $\alpha = 0$, in accordance with the asymptotic behaviour (6.30) that all aperture-limited functions must possess.

The compromise between beamwidth and sidelobe ratio that is the object of the Taylor design procedure can be seen directly in Fig. 7.2: suppressing the sidelobes by increasing the value of the design sidelobe ratio cosh πA causes the beamwidth between first nulls to increase quite noticeably.

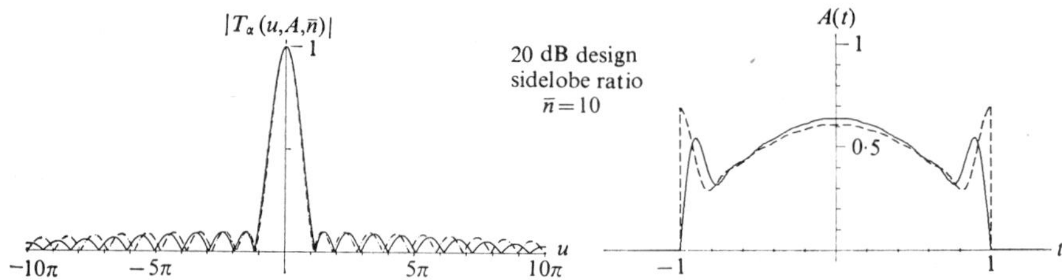

FIG. 7.3. The Taylor pattern space factor $T_\alpha(u, A, \bar{n})$ and its aperture distribution $A(t)$ for $\alpha = 0$ (broken) and $\alpha = 1$ (solid) for $\bar{n} = 10$ with a design sidelobe ratio of 20 dB.

Since the pattern functions are all normalized to a common value (unity) in the broadside direction $u = 0$, the relative magnitudes of the aperture distributions required for these five cases can be compared directly to obtain a rough indication of the relative amounts of observable stored energy, and hence of the relative values of Q, needed to obtain the pattern approximation. Such a comparison gives an indication of the relative feasibility of various source synthesis solutions, regardless of what particular criterion for pattern goodness might be chosen, for whenever the accuracy of the pattern approximation is forced to become too great—as in the case of superdirective distributions that will be treated later in Chapter 8, for example—the Q of the antenna can increase dramatically. And any increase in Q is invariably accompanied by an increase in the magnitude of the aperture distribution. In all of the five cases shown, however, the aperture distributions for both $\alpha = 0$ and 1 are seen to lie remarkably near the value 0·5 of the uniform distribution, the latter of which will be seen in Chapter 8 to have nearly the lowest value of Q possible for any given size of aperture. Hence it is clear that none of these five cases should be difficult to realize physically.

By choosing the value of the parameter $\bar{n}$ to be sufficiently large the general Taylor pattern for any value of α can be made to approximate the ideal van der Maas pattern as closely as desired over any finite region about the origin. If the size of the central region defined by $n \leqslant \bar{n}$ were allowed to increase indefinitely the Taylor pattern would approach the van der Maas pattern itself and, hence, the Q of the antenna would grow without bound. But comparison of the 20 dB sidelobe ratio designs for the two cases $\bar{n} = 5$ and 10 indicates that the rate at which the value of Q increases with increasing values of $\bar{n}$ must be rather slow.

From these numerical results it is concluded that all practical differences in operating performance between Taylor's original distribution for $\alpha = 0$ and the new one for $\alpha = 1$ can be made negligibly small by choosing the value of $\bar{n}$ to be sufficiently large. The Taylor patterns for $\alpha = 0$ and 1 would both approach the ideal van der Maas pattern. And, since the value of Q is expected to be essentially the same for both $\alpha = 0$ and 1, the best design for both cases would be that for which $\bar{n}$ is chosen to have its largest practicable value. Even for a value no larger than $\bar{n} = 10$ it is evident from Fig. 7.3 that the essential features of both are becoming very much alike.

CHAPTER 8
MAXIMUM DIRECTIVITY[8]

ONE of the most celebrated problems in the history of antenna theory is to find the aperture distribution that produces maximum directivity for a continuous line source of given length. It rose to prominence with the publication of the famous paper by Bouwkamp and de Bruijn† in which they proved that a solution published earlier by La Paz and Miller‡ was incorrect and that, in fact, there is no theoretical limit to the directivity available. This was confirmed independently by Woodward and Lawson in their treatment of strip sources. Woodward and Lawson showed, in addition, that the well-known failure to achieve in practice any directivity much in excess of that of a uniform distribution occurs because of a sharp increase in the net reactive power that would be required at the source to produce it. It is now known (section 5.4i) that the net reactive power for Woodward and Lawson's strip source is equal to the total reactive power observable at the source. To be realizable physically, then, some constraint should be imposed upon the proportion of reactive to radiative power, or, equivalently, upon the quality factor Q. The maximum directivity problem thus becomes one of optimizing the relationship between directivity and Q.

A class of optimum solutions to the maximum directivity problem will be developed here for strip and line sources. It is optimum in the sense that it results in the maximum possible value of a generalization D_α of the directivity of a strip or a line source of length l when constrained to have a specified value of γ_α for any value of α greater than -1. For $\alpha = -\frac{1}{2}, \frac{1}{2}$, and 1 the quantity D_α will be seen to represent true directivity for E- and H-plane strip sources and for H-plane line sources, respectively. And the constraining parameter γ_α has already been seen to represent either an exact or an approximate measure of Q. Thus, this one class of optimum solutions applies to all of the three linear sources of physical interest.

By allowing the value of γ_α to become arbitrarily large this class of optimum synthesis solutions leads to an arbitrarily high value for the maximum directivity $D_{\alpha\max}$ available from an aperture of any given length, in full agreement with the conclusion reached earlier by Bouwkamp and de Bruijn.

† C. J. Bouwkamp and N. G. de Bruijn, *loc. cit.*

‡ L. La Paz and G. A. Miller. Optimum current distributions on vertical antennas. *Proc. Inst. Radio Engrs*, **31**, pp. 214–32, May 1943.

8.1. Directivity of strip and line sources

In contrast to resolution, for which no precise measure exists, there is a very precise measure for directivity. It is defined as the ratio of power radiated per unit angle (plane angle for strip sources, solid angle for line sources) in the direction of maximum radiation to the average power radiated per unit angle. Thus, it is solely a measure of the extent to which an antenna concentrates its radiated power in the neighbourhood of a given direction, without regard for height of the sidelobes. In the case of strip and line sources the formulation for directivity becomes particularly simple because the two-dimensional radiation pattern reduces to a one-dimensional function. The formulation differs somewhat for the three different cases of E-plane strip sources, H-plane strip sources, and H-plane line sources, however, so each will be treated separately.

(i) *E-plane strip sources*

The radiated power of E-plane strip sources is concentrated in the $\phi = 0$ plane, where its density is proportional to $|E_\theta|^2$. Hence the power density is proportional to $|F_{xx}(c_x \sin\theta)|^2$. The directivity of E-plane strip sources in any given direction $\theta = \theta_0$ is then

$$D = \frac{|F_{xx}(c_x \sin\theta_0)|^2}{\dfrac{1}{\pi}\displaystyle\int_{-\pi/2}^{\pi/2} |F_{xx}(c_x \sin\theta)|^2\, d\theta} = \frac{\pi\, |F_{xx}(c_x \eta_0)|^2}{\displaystyle\int_{-1}^{1} |F_{xx}(c_x \eta)|^2 (1-\eta^2)^{-\frac{1}{2}}\, d\eta}, \tag{8.1}$$

in which η denotes the direction-cosine $\sin\theta$.

(ii) *H-plane strip sources*

For H-plane strip sources the radiated power is concentrated in the $\phi = \pi/2$ plane, its density being proportional to $|E_\phi|^2$. Hence the power density is proportional to $|\cos\theta F_{xy}(c_y \sin\theta)|^2$. The directivity of H-plane strip sources in any given direction $\theta = \theta_0$ is then

$$D = \frac{|\cos\theta_0 F_{xy}(c_y \sin\theta_0)|^2}{\dfrac{1}{\pi}\displaystyle\int_{-\pi/2}^{\pi/2} |\cos\theta F_{xy}(c_y \sin\theta)|^2\, d\theta} = \frac{\pi\, |F_{xy}(c_y \eta_0)|^2 (1-\eta_0^2)}{\displaystyle\int_{-1}^{1} |F_{xy}(c_y \eta)|^2 (1-\eta^2)^{\frac{1}{2}}\, d\eta}, \tag{8.2}$$

in which η again denotes the direction-cosine $\sin\theta$.

(iii) *H-plane line sources*

In the case of H-plane line sources the radiated power is spread out uniformly in all directions about the axis of the source. If the polar axis of the spherical coordinate system is taken to be the y-axis on which the line source is located (instead of the customary z-axis), as indicated in Fig. 8.1, then the radiated power density will be proportional to $|E_{\phi_y}|^2$, which is proportional to $|\sin\theta_y F_{xy}(c_y \cos\theta_y)|^2$. Hence the directivity of H-plane line sources in any given direction θ_{y0} is

$$D = \frac{|\sin\theta_{y0}F_{xy}(c_y\cos\theta_{y0})|^2}{\frac{1}{4\pi}\int\limits_{\phi_y=0}^{2\pi}\int\limits_{\theta_y=0}^{\pi}|\sin\theta_y F_{xy}(c_y\cos\theta_y)|^2\sin\theta_y\,d\phi_y\,d\theta_y}$$

$$= \frac{2\,|F_{xy}(c_y\eta_0)|^2\,(1-\eta_0^2)}{\int\limits_{-1}^{1}|F_{xy}(c_y\eta)|^2\,(1-\eta^2)\,d\eta}, \tag{8.3}$$

where η denotes the direction-cosine $\cos\theta_y$.

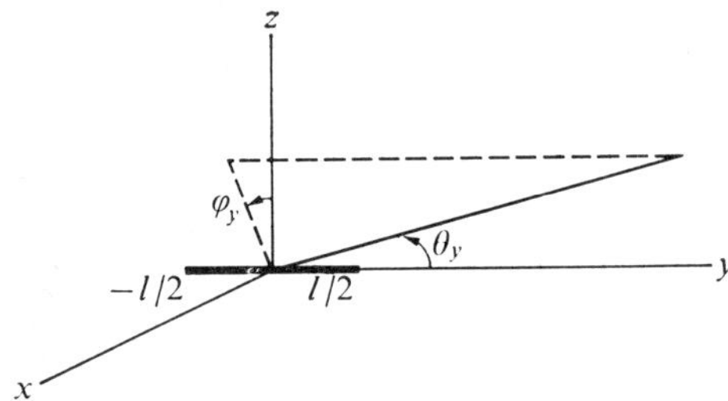

FIG. 8.1. Coordinate system for a line source of length l.

Although these three expressions for directivity are all different there is a certain similarity among them, particularly if the direction of maximum radiation for the two strip sources is restricted to the broadside direction $\theta_0 = 0$. In the latter event all three become special cases of the following *generalized directivity*,

$$D_\alpha^\beta = \frac{K\,|F_\alpha(c\eta_0)|^2\,(1-\eta_0^2)^\beta}{\int\limits_{-1}^{1}|F_\alpha(c\eta)|^2\,(1-\eta^2)^\beta\,d\eta}, \tag{8.4}$$

in which the parameter c denotes π times the electrical length l/λ of the aperture and the constant K has the value π for strip sources and 2 for line sources, where β is $-\frac{1}{2}$ for E-plane strip sources, $\frac{1}{2}$ for H-plane strip sources, and 1 for H-plane line sources.

The expressions given above for the directivity of strip and line sources are exact. They are frequently approximated for highly directive pattern space factors $F_\alpha(c\eta)$, however, by omitting the weight factor $(1-\eta^2)^\beta$. That would correspond to a fourth special case of the generalized directivity, namely, the case in which $\beta = 0$.

It should be noted that there is a striking resemblance between this generalization (8.4) of the directivity of strip and line sources and the generalization γ_α^β of Taylor's γ that was formulated in (5.34) as an alternative to Q. Both generalizations have exactly the value required in the denominator for each of the three cases when the value of β is chosen to be $-\frac{1}{2}$, $\frac{1}{2}$, or 1. And both have either exactly or approximately the value required in the numerator. But both also have the disadvantage of being too general to fit into the mathematical framework of the characteristic spheroidal functions. In order to take full advantage of the properties peculiar to that framework it will again be necessary to assume that $\beta = \alpha$. The generalized directivity D_α^β then becomes what will be called the *directivity parameter* D_α,

$$D_\alpha = \frac{K\,|F_\alpha(c\eta_0)|^2\,(1-\eta_0^{\,2})^\alpha}{\int\limits_{-1}^{1} |F_\alpha(c\eta)|^2\,(1-\eta^2)^\alpha\,d\eta}. \tag{8.5}$$

This is the quantity that will be maximized here. It corresponds to D_α^β in precisely the same way that the constraining parameter γ_α corresponds to γ_α^β.

Thus, a good approximation to the synthesis problem of optimizing the relationship between directivity and Q for strip and line sources can be expected by optimizing, instead, the relationship between D_α and γ_α. When the value of α is chosen to be $-\frac{1}{2}$, $\frac{1}{2}$, or 1 this optimum relationship between D_α and γ_α represents the optimum Q-constrained solution for E- and H-plane strip sources and H-plane line sources, respectively. And when the value of α is chosen to be 0 it represents the optimum solution obtained for the usual approximation to directivity when constrained by Taylor's superdirectivity ratio γ.

The importance of this particular formulation of the maximum directivity problem lies in the fact that it can be solved exactly in terms of the characteristic spheroidal functions $\{\psi_{\alpha n}(c, \eta)\}$ by making use of their double orthogonality properties. That solution will be developed in detail in section 8.3.

8.2. Uniform distribution

Experimental evidence indicates that it is difficult to obtain values for directivity that are much in excess of that produced by an aperture distribution which is uniform in both magnitude and phase. For that reason the uniform distribution is frequently regarded as having the largest directivity that can be achieved in practice for an aperture of given length. Any aperture distribution that would give more directivity than the uniform distribution is usually said to be 'superdirective'.

This qualitative observation regarding the maximum directivity achievable in practice will now be given a more precise quantitative meaning by showing that the uniform distribution does, indeed, produce maximum directivity, although in a somewhat more restrictive sense than is usually attributed to it. As might be expected, the sense in which the directivity is maximum must be restricted not only by the length of the aperture but also by some form of constraint on the proportion of reactive to radiative power. The constraining parameter turns out to be Taylor's superdirectivity ratio γ, the special case of γ_α in which the value of the edge exponent α is zero (the only value possible for a uniform distribution).

The expression for directivity will be taken to be the parameter D_α for $\alpha = 0$. This is admittedly not the true directivity for either a strip or a line source but it is the approximation most often used. And it is the only expression for directivity that is known to lead to the uniform distribution as an extremum.

Assuming that maximum radiation is to occur in the broadside direction $\eta_0 = 0$, the ratio of D_0 to γ_0 for an arbitrary aperture distribution $f(t)$ (the edge factor is always unity for $\alpha = 0$) is seen to be

$$\frac{D_0}{\gamma_0} = \frac{K\,|F_0(0)|^2}{\frac{1}{c}\int\limits_{-\infty}^{\infty} |F_0(u)|^2\,du} = \frac{cK}{2\pi}\frac{\left|\int\limits_{-1}^{1} f(t)\,dt\right|^2}{\int\limits_{-1}^{1} |f(t)|^2\,dt}, \tag{8.6}$$

the latter being obtained by substituting the integral representation (5.9) for $F_0(u)$. Now suppose that $f(t)$ is restricted to the class of functions for which γ_0 is assigned the particular value γ_{unif} of the uniform distribution,

$$\gamma_{\text{unif}} = \frac{\pi/2}{\text{Si}\,2c-(\sin^2 c)/c}, \tag{8.7}$$

obtained from (5.35). This is a wide class of functions that includes among them the uniform distribution itself as but one special case. Within this class the particular function that maximizes the ratio (8.6) follows directly from Schwartz's inequality,

$$\left|\int\limits_{-1}^{1} f(t)\,dt\right|^2 \leqslant 2\int\limits_{-1}^{1} |f(t)|^2\,dt, \tag{8.8}$$

in which equality is obtained when, and only when, $f(t)$ is any complex constant. Thus, when γ_0 is restricted to have the value γ_{unif} the maximum value of the ratio (8.6) is seen to be

$$\left(\frac{D_0}{\gamma_{\text{unif}}}\right)_{\max} = \frac{cK}{\pi}, \tag{8.9}$$

which occurs when, and only when, the aperture distribution $f(t)$ is uniform in both magnitude and phase. Since the maximum value of the ratio D_0/γ_{unif} must give the maximum value of the directivity D_0 that can be obtained when γ_0 is chosen to have the particular value γ_{unif}, it is concluded that *the uniform distribution results in maximum directivity in the broadside direction when, and only when, the value of γ_0 is constrained to be precisely γ_{unif}*. It is in this sense, and only in this sense, that the uniform distribution is more directive than any other aperture distribution of given length. For values of γ_0 that differ from γ_{unif} there will still be an aperture distribution that results in maximum directivity, as will be shown next, but it will differ from the uniform distribution.

8.3. Optimum solution

A more general solution of the maximum directivity problem will now be obtained in which arbitrary values of $\alpha > -1$, of the length parameter $c = \pi l/\lambda$, of the constraining parameter $\gamma_\alpha \geqslant \gamma_{\alpha 0}(c)$, and of the direction cosine η_0 of maximum radiation can be specified as independent design parameters. It will be seen that the simple solution obtained above in section 8.2 for the uniform distribution emerges as an interesting special case.

The objective of the maximum directivity problem as formulated herein is to find the aperture distribution whose pattern space factor maximizes expression (8.5) for the directivity parameter D_α under the constraining condition that the defining equation (5.35) for γ_α is to be satisfied for an arbitrarily prescribed value of γ_α that is no less than its absolute minimum value $\gamma_{\alpha 0}(c)$. Thus, the pattern space factor required will be the aperture-limited function that maximizes the constrained directivity functional

$$D_\alpha[F_\alpha] = \frac{K\,|F_\alpha(c\eta_0)|^2\,(1-\eta_0{}^2)^\alpha}{\int\limits_{-1}^{1} |F_\alpha(c\eta)|^2\,(1-\eta^2)^\alpha\,d\eta} + \mu\left\{\frac{\int\limits_{-\infty}^{\infty} |F_\alpha(c\eta)|^2\,|1-\eta^2|^\alpha\,d\eta}{\int\limits_{-1}^{1} |F_\alpha(c\eta)|^2\,(1-\eta^2)^\alpha\,d\eta} - \gamma_\alpha\right\}, \tag{8.10}$$

and that simultaneously satisfies (5.35) as a constraining side condition. The parameter μ is a Lagrange multiplier whose value is yet to be determined.

The maximization problem will be recast into an equivalent form by expanding the pattern space factor in an orthogonal series of aperture-limited basis functions. Instead of seeking a continuous function $F_\alpha(c\eta)$ the problem then becomes one of seeking a discrete set of expansion coefficients. Any set of functions that is orthogonal and complete with respect to all aperture-limited functions on $(-\infty, \infty)$ would do, in principle. But the maximization problem would simplify greatly if the set were orthogonal over the visible region $(-1, 1)$ as well. This double orthogonality property, which is the key to the optimum solution that will be developed here and to the optimum solution of the pattern shaping problem that will be developed in Chapter 9, is possessed only by the spheroidal functions $\{\psi_{\alpha n}(c, \eta)\}$. Hence the pattern space factor will be expressed as an $N+1$-term series of spheroidal functions of order α,

$$F_\alpha(c\eta) = \sum_{n=0}^{N} a_n \psi_{\alpha n}(c, \eta), \tag{8.11}$$

with unknown expansion coefficients $\{a_n\}$. By choosing N sufficiently large the series can be made to approximate any aperture-limited pattern space factor as closely as desired over $(-\infty, \infty)$, from the completeness property established in Appendix 3.

With this representation for $F_\alpha(c\eta)$ the constrained directivity functional (8.10) then becomes

$$D_\alpha[a_0, a_1, \ldots, a_N] = \frac{K \sum_{n=0}^{N} \sum_{m=0}^{N} a_n a_m^* \psi_{\alpha n}(c, \eta_0) \psi_{\alpha m}(c, \eta_0)(1-\eta_0^2)^\alpha + \mu \sum_{n=0}^{N} |a_n|^2 \{\gamma_{\alpha n}(c) - \gamma_\alpha\} \Lambda_{\alpha n}(c),}{\sum_{n=0}^{N} |a_n|^2 \Lambda_{\alpha n}(c)} \tag{8.12}$$

after making use of the double orthogonality relations (5.12) of the spheroidal functions.

The optimization problem now becomes one of finding the $N+1$ unknown coefficients and the undetermined Lagrange multiplier μ such that the constrained directivity $D_\alpha[a_0, a_1, \ldots, a_N]$ takes its largest possible value, under the side condition that the set of a_ns is to be constrained to satisfy

$$\sum_{n=0}^{N} |a_n|^2 \{\gamma_{\alpha n}(c) - \gamma_\alpha\} \Lambda_{\alpha n}(c) = 0, \tag{8.13}$$

obtained from (5.35) after again using the orthogonality relations (5.12). A necessary condition for an extremum is that every partial derivative $\partial D/\partial a_p^*$ be zero. Multiplying both sides of (8.12) by the denominator on the right-hand side, then differentiating through with respect to each a_p^* (note that a_p and its complex conjugate a_p^* are linearly independent), and finally setting each partial derivative equal to zero, a set of $N+1$ linear algebraic equations for the $N+1$ unknown a_ns is obtained,

$$\sum_{n=0}^{N} \{K \psi_{\alpha n}(c, \eta_0) \psi_{\alpha p}(c, \eta_0)(1-\eta_0^2)^\alpha + [\mu\{\gamma_{\alpha n}(c) - \gamma_\alpha\} - D_\alpha] \Lambda_{\alpha n}(c)\, \delta_{pn}\} a_n = 0,$$

$$p = 0, 1, 2, \ldots, N, \tag{8.14}$$

in which the diagonal coefficients of the a_ns contain two unknown quantities μ and D_α. For these equations to be consistent the determinant of coefficients must vanish

$$|\{x_{pn}(\mu, D_\alpha)\}| = \begin{vmatrix} x_{00}(\mu, D_\alpha) & x_{01} & x_{02} & \cdots & x_{0N} \\ x_{10} & x_{11}(\mu, D_\alpha) & x_{12} & \cdots & x_{1N} \\ \cdot & & \cdot & & \cdot \\ \cdot & & & \cdot & \cdot \\ \cdot & & & & \cdot \\ x_{N0} & \cdots & \cdots & \cdots & x_{NN}(\mu, D_\alpha) \end{vmatrix} = 0, \tag{8.15}$$

wherein the individual elements are defined to be

$$x_{pn}(\mu, D_\alpha) = K\psi_{\alpha n}(c, \eta_0)\psi_{\alpha p}(c, \eta_0)(1-\eta_0^2)^\alpha + \\ +[\mu\{\gamma_{\alpha n}(c)-\gamma_\alpha\}-D_\alpha]\Lambda_{\alpha n}(c)\,\delta_{pn}. \tag{8.16}$$

The set of simultaneous equations (8.14) is now overdetermined, hence one of the a_ns must be arbitrary. Let that one be a_0. Furthermore, one of the equations is now redundant and can be removed. Let it be the first one ($p = 0$). The remaining N equations in (8.14) can then be written

$$\sum_{n=1}^{N} x_{pn}(\mu, D_\alpha)a_n = -x_{p0}a_0, \qquad p = 1, 2, 3, \dots, N. \tag{8.17}$$

They can be solved simultaneously for the N unknown a_ns in terms of the arbitrarily chosen value of a_0 and the as yet undetermined values of μ and D_α. These N equations, together with the determinantal condition (8.15) and the constraining side condition (8.13), must then determine all of the $N+2$ unknowns.

The above deduction that one of the a_ns must be arbitrary is completely consistent with the fact that the maximum directivity criterion says nothing whatsoever about the choice of scale factor for the aperture or pattern functions. Either the aperture or the pattern function can be normalized at will to have any scale factor desired, real or complex. By choosing the value of a_0 to be real all of the other a_ns will also be real, since the x_{pn}s are all real. Thus, by choosing a_0 real, *both the aperture and pattern space factors required for maximum directivity will always be purely real.*

Numerical solutions can be obtained by choosing some approximate value for μ and D_α and then computing successive corrections. First define two functions $g(\mu, D_\alpha)$ and $h(\mu, D_\alpha)$, as follows:

$$g(\mu, D_\alpha) = |\{x_{pn}(\mu, D_\alpha)\}|, \tag{8.18}$$

$$h(\mu, D_\alpha) = a_0^2\{\gamma_{\alpha 0}(c)-\gamma_\alpha\}\Lambda_{\alpha 0}(c) + \sum_{n=1}^{N} a_n^2(\mu, D_\alpha)\{\gamma_{\alpha n}(c)-\gamma_\alpha\}\Lambda_{\alpha n}(c), \tag{8.19}$$

where $a_n(\mu, D_\alpha)$ denotes the dependence of the expansion coefficient a_n on μ and D_α. For the correct value of μ and D_α that maximizes (8.12),

$$g(\mu, D_\alpha) = 0, \tag{8.20}$$

$$h(\mu, D_\alpha) = 0, \tag{8.21}$$

the first being the determinantal equation (8.15) and the second the constraining condition (8.13), since the a_ns are all real. For any approximate value μ_a and $D_{\alpha a}$, however, these will be in error by an amount Δg and Δh, respectively, such that

$$g(\mu_a, D_{\alpha a})+\Delta g(\mu_a, D_{\alpha a}) = 0, \tag{8.22}$$

$$h(\mu_a, D_{\alpha a})+\Delta h(\mu_a, D_{\alpha a}) = 0. \tag{8.23}$$

To the first order this error will be

$$\Delta g = \frac{\partial g}{\partial \mu}\Delta\mu_a+\frac{\partial g}{\partial D_\alpha}\Delta D_{\alpha a} = -g, \tag{8.24}$$

$$\Delta h = \frac{\partial h}{\partial \mu}\Delta\mu_a+\frac{\partial h}{\partial D_\alpha}\Delta D_{\alpha a} = -h. \tag{8.25}$$

Solving simultaneously for the first-order corrections to μ_a and $D_{\alpha a}$,

$$\Delta\mu_\alpha = -\frac{g(\partial h/\partial D_\alpha)-h(\partial g/\partial D_\alpha)}{(\partial g/\partial\mu)(\partial h/\partial D_\alpha)-(\partial h/\partial\mu)(\partial g/\partial D_\alpha)}, \tag{8.26}$$

$$\Delta D_{\alpha a} = -\frac{h(\partial g/\partial\mu)-g(\partial h/\partial\mu)}{(\partial g/\partial\mu)(\partial h/\partial D_\alpha)-(\partial h/\partial\mu)(\partial g/\partial D_\alpha)}. \tag{8.27}$$

To obtain the partial derivatives $\partial g/\partial\mu$ and $\partial g/\partial D_\alpha$ note that only the diagonal elements x_{nn} contain μ and D_α. Hence, from the chain rule for partial derivatives one finds that

$$\frac{\partial g}{\partial\mu} = \sum_{n=0}^{N}\frac{\partial g}{\partial x_{nn}}\frac{\partial x_{nn}}{\partial\mu} = \sum_{n=0}^{N} A_{nn}\{\gamma_{\alpha n}-\gamma_\alpha\}\Lambda_{\alpha n}, \tag{8.28}$$

$$\frac{\partial g}{\partial D_\alpha} = \sum_{n=0}^{N}\frac{\partial g}{\partial x_{nn}}\frac{\partial x_{nn}}{\partial D_\alpha} = -\sum_{n=0}^{N} A_{nn}\Lambda_{\alpha n}, \tag{8.29}$$

where A_{nn} denotes the minor of element x_{nn}. The other partials $\partial h/\partial\mu$ and $\partial h/\partial D_\alpha$ are obtained directly by differentiation of (8.19),

$$\frac{\partial h}{\partial\mu} = 2\sum_{n=1}^{N} a_n\frac{\partial a_n}{\partial\mu}\{\gamma_{\alpha n}-\gamma_\alpha\}\Lambda_{\alpha n}, \tag{8.30}$$

$$\frac{\partial h}{\partial D_\alpha} = 2\sum_{n=1}^{N} a_n\frac{\partial a_n}{\partial D_\alpha}\{\gamma_{\alpha n}-\gamma_\alpha\}\Lambda_{\alpha n}, \tag{8.31}$$

where $\partial a_n/\partial \mu$ and $\partial a_n/\partial D_\alpha$ are obtained by first differentiating (8.17) partially with respect to μ and D_α, respectively,

$$\sum_{n=1}^{N} x_{pn}\frac{\partial a_n}{\partial \mu} = -\{\gamma_{\alpha p}-\gamma_\alpha\}\Lambda_{\alpha p}a_p, \qquad p = 1, 2, 3, \ldots, N, \quad (8.32)$$

$$\sum_{n=1}^{N} x_{pn}\frac{\partial a_n}{\partial D_\alpha} = \Lambda_{\alpha p}a_p, \qquad p = 1, 2, 3, \ldots, N, \quad (8.33)$$

and then solving simultaneously.

The pattern space factor for maximum directivity can now be computed from its series expansion (8.11) by using the expansion coefficients $\{a_n\}$ determined from this maximization process. The aperture distribution producing it, which is the optimum solution sought, can also be computed from these same expansion coefficients by

$$\begin{aligned} A(t) &= (1-t^2)^\alpha f(t) \\ &= (1-t^2)^\alpha \sum_{n=0}^{N} \frac{a_n}{\nu_{\alpha n}(c)}\psi_{\alpha n}(c, t), \qquad -1 \leqslant t \leqslant 1. \end{aligned} \quad (8.34)$$

This is easily verified by inserting the series into (5.9) and integrating term by term. The integral relation (5.11) reduces the resulting representation for $F_\alpha(c\eta)$ to precisely the series (8.11).

The only part of the problem that still remains unspecified is the procedure for obtaining a starting approximation for μ and D_α that is close enough to the correct value to ensure that the process will converge to the absolute maximum value of D_α and not to some other extremum. The procedure used to compute the results that will be shown here is based upon the *a priori* knowledge (developed in section 5.6) that an absolute lower bound exists on the value of γ_α, namely, $\gamma_{\alpha 0}(c)$, and that there is one and only one pattern space factor for that lower bound, namely, the spheroidal function $\psi_{\alpha 0}(c, \eta)$. Consequently there exists an absolute lower bound on the maximum directivity obtainable in any given direction, the absolute lower bound being that produced by the spheroidal function $\psi_{\alpha 0}(c, \eta)$. In the broadside direction $\eta = 0$ it is

$$D_{\alpha\max}\Big|_{\gamma_\alpha=\gamma_{\alpha 0}(c)} = \frac{2\psi_{\alpha 0}^2(c, 0)}{\Lambda_{\alpha 0}(c)}; \quad (8.35)$$

i.e. for a pattern space factor represented by the series (8.11) it is the directivity produced in the broadside direction by just the first term alone. For some value of γ_α slightly greater than the lower bound $\gamma_{\alpha 0}(c)$ the value of maximum directivity should be slightly greater than its lower bound (8.35), determined largely, but not completely, by the first term. A good approximation should then be obtainable by retaining just the first two terms of the series. In the case of the computations that will be shown here for $\alpha = 0$ this assumption leads to two roots for μ and D_α, only one of which provides a possible approximation to the correct extremum because its value of $D_{\alpha\max}$ is larger than (8.35) while the other is smaller and, hence, impossible. The exact value of μ and $D_{\alpha\max}$ obtained by means of successive iterations from this approximation can then be used as the approximate value for a third, slightly larger, value of γ_α, and so on.

By this process of carefully extrapolating upward in γ_α from its lower bound $\gamma_{\alpha 0}(c)$ the maximum directivity $D_{\alpha\max}$ in the broadside direction of a

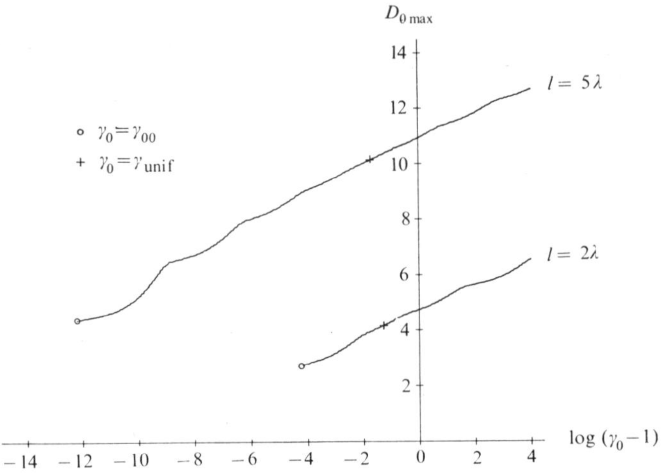

FIG. 8.2. The maximum possible directivity $D_{0\max}$ in the broadside direction of a line source of length l as a function of $\log(\gamma_0-1)$ for $l = 2\lambda$ and 5λ. The circle indicates the value of $D_{0\max}$ at the absolute minimum value $\gamma_{00}(c)$ of γ_0, and the cross indicates the value of $D_{0\max}$ at the value γ_{unif} of the uniform distribution.

line source ($K = 2$) has been computed for $\alpha = 0$ as a function of $\log(\gamma_0-1)$ for two different aperture lengths $l = 2\lambda$ and 5λ. The value $\alpha = 0$ was chosen in order to be able to check the value of $D_{0\max}$ obtained independently in section 8.2 for the uniform distribution. The resulting graphs are shown in Fig. 8.2 for all values of γ_0 from its lower bound $\gamma_{00}(c)$, indicated by the circles, up to a value of 10 000. Both graphs are seen to be roughly linear, with a slope of slightly less than 0·5. From this slope it is concluded that for $\alpha = 0$ *the maximum directivity available from a line source of any given length will increase by roughly one unit for every hundredfold increase in the value of* γ_0-1. This is a vivid quantitative indication of the enormous rate of growth of the effects of energy storage in the fields of line sources.

Similar results would be obtained for strip sources in the broadside direction. The only difference would be an increase in the scale of $D_{0\max}$ by a factor of $\frac{1}{2}\pi$ in Fig. 8.2, since the value of the factor K changes from 2 to π.

Since the uniform distribution was found in section 8.2 to give the maximum possible value of the directivity D_0 when the value of γ_0 is constrained to be γ_{unif}, its directivity for a line source, given by

$$D_{\text{unif}} = \frac{2c}{\pi}\gamma_{\text{unif}} \tag{8.36}$$

from (8.9) for $K = 2$, is expected to appear as a point on each of the graphs in Fig. 8.2. When this value (8.36) was compared numerically with the value obtained quite independently from the optimum synthesis solution by use of the spheroidal function expansion the two were found to agree exactly, to 15 out of 16 significant figures, for both of the aperture lengths used. It is indicated by the crosses shown on the graphs in Fig. 8.2. This is convincing evidence of the validity of the synthesis solution obtained here and of the accuracy of the numerical results shown. It also provides, incidentally, an alternative starting approximation that can be used for the values of μ and D_α in the particular case of $\alpha = 0$: When γ_0 is chosen to have the particular value γ_{unif} obtained from (8.7) the maximum value of D_0 must be precisely that given by (8.36). The corresponding value of μ has been observed numerically to be

$$\mu_{\text{unif}} = -c/\pi, \tag{8.37}$$

although no analytical reason for it is known as yet.

This constitutes a complete solution to the maximum directivity problem. From a pair of approximate values of μ and D_α successively better approximations can be obtained by adding to them the corrections $\Delta\mu_a$ and $\Delta D_{\alpha a}$ computed from (8.26) and (8.27). The corrections involve the a_ns and their partial derivatives computed by solving simultaneously the individual sets of linear equations (8.17), (8.32), and (8.33). By using reasonably accurate starting values for μ and D_α the process has been found to converge quickly to the correct values. It results in the best approximation to the absolute maximum value of directivity D_α, and to the corresponding values of the expansion coefficients $\{a_n\}$ that determine the aperture and pattern space factors, that can be achieved with $N+1$ terms for a given value of the constraining parameter γ_α. And the accuracy of the approximation can be made arbitrarily precise by choosing the value of N to be sufficiently large. In practice the value of N need not be much larger than $2l/\lambda$ because of the fact that the series representing the constrained solution always converges rapidly after $(2l/\lambda)+1$ terms.

8.4. Illustration for $\alpha = 0$

The optimum solution derived above will be illustrated here by showing the computed pattern space factor $F_0(c\eta)$ and its aperture distribution

11

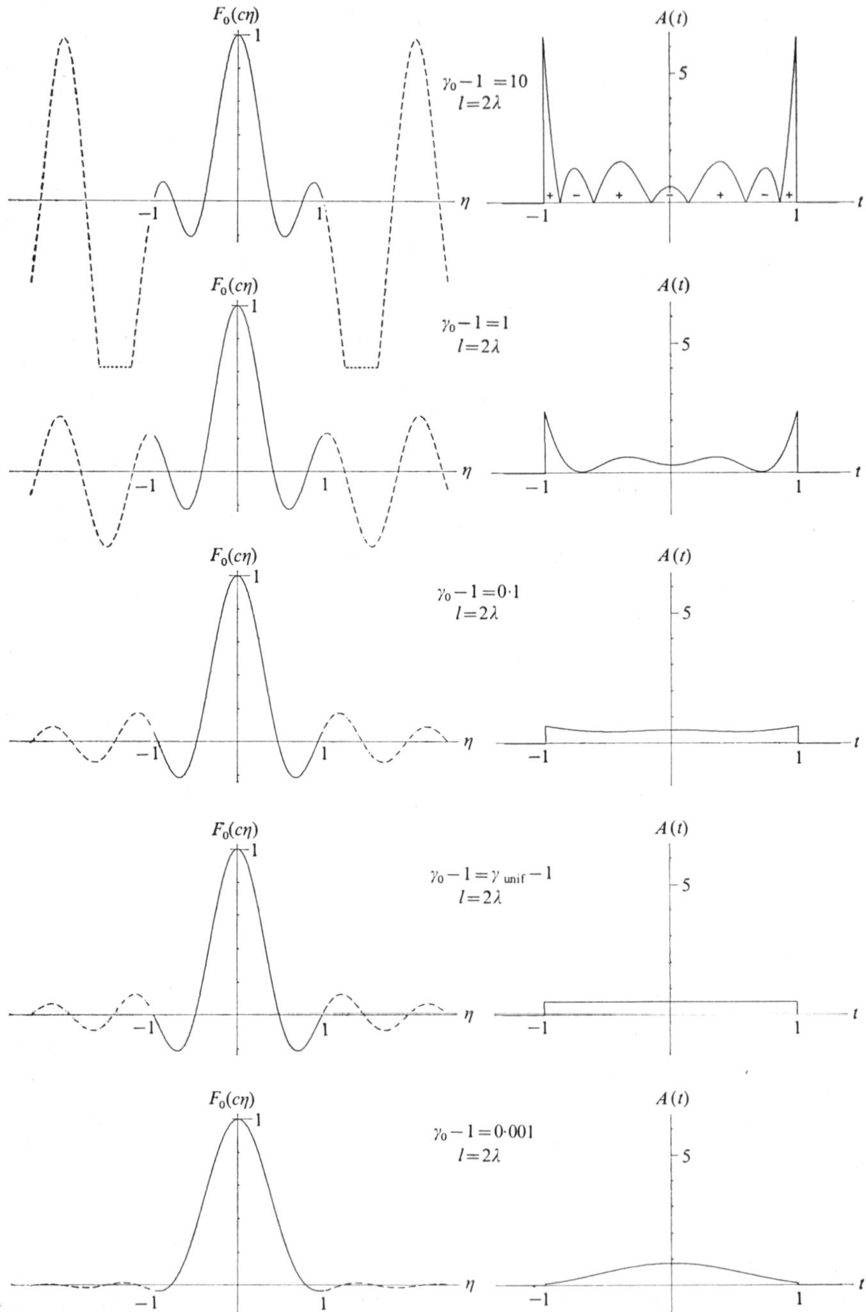

FIG. 8.3. The pattern space factor $F_0(c\eta)$ and its aperture distribution $A(t)$ for maximum directivity in the broadside direction $\eta_0 = 0$ when constrained by $\gamma_0-1 = 10$, 1, 0·1, $\gamma_{\text{unit}}-1$, and 0·001, for $l = 2\lambda$.

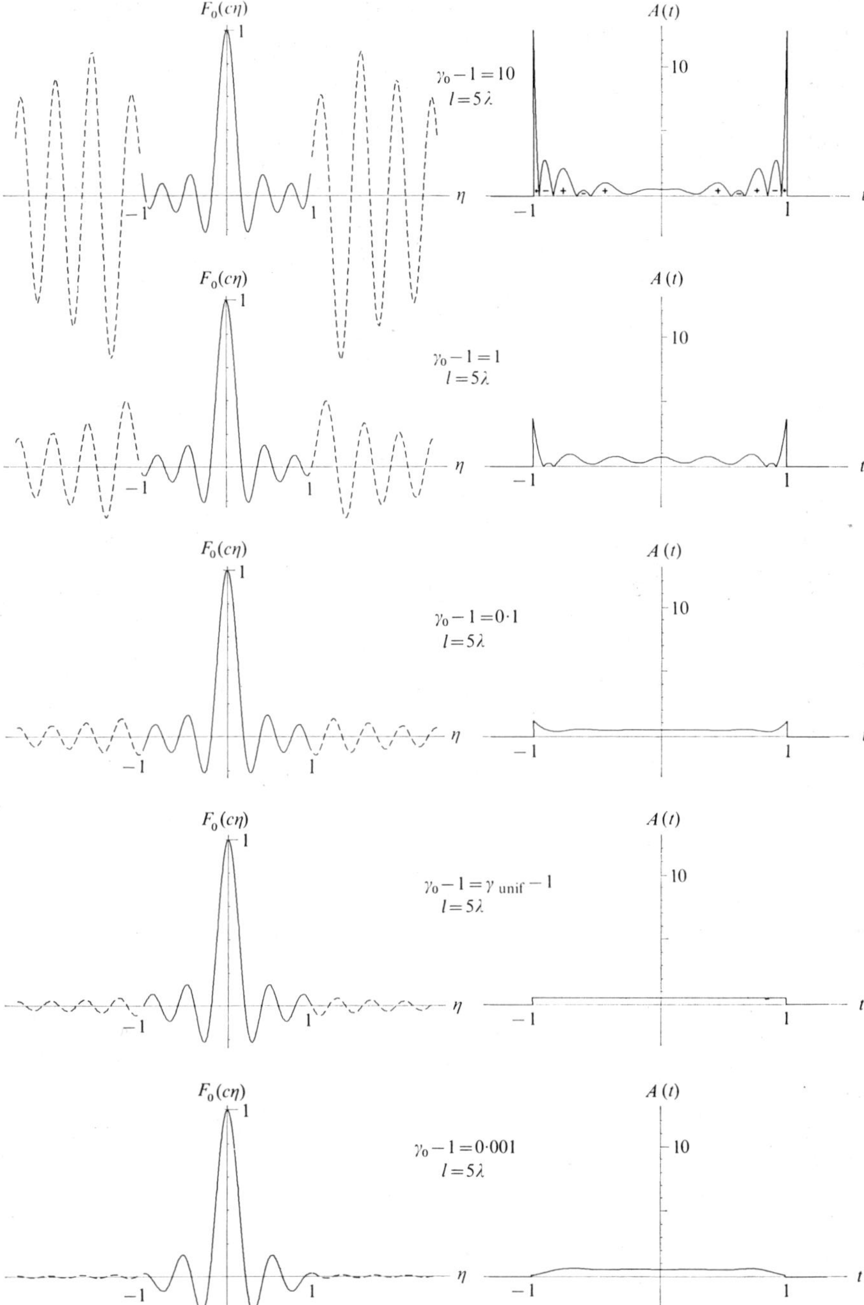

FIG. 8.4. The pattern space factor $F_0(c\eta)$ and its aperture distribution $A(t)$ for maximum directivity in the broadside direction $\eta_0 = 0$ when constrained by $\gamma_0-1 = 10$, 1, 0·1, $\gamma_{\text{unif}}-1$, and 0·001, for $l = 5\lambda$.

$A(t)$ for various values of the aperture length and of the constraining parameter γ_0. The value of the edge exponent α is taken to be zero in order to be able to include the special case of the uniform distribution that was treated independently in section 8.2.

The pattern space factor (8.11) and its aperture distribution (8.34) for maximum directivity in the broadside direction $\eta_0 = 0$ are shown for $\alpha = 0$ in Figs. 8.3 and 8.4 for aperture lengths of $l = 2\lambda$ and 5λ, respectively, for each of five different values of the constraining parameter γ_0. The corresponding values of μ and $D_{0\max}$ are shown in Table 8.1 (all computations were

TABLE 8.1

$D_{0\max}$ and μ

$\gamma_0 - 1$	$D_{0\max}$	μ
	$l = 2\lambda$	
10	0·5250484805461806+01	−0·1317156193857867−01
1	0·4785896183301008+01	−0·7487493127704290−01
0·1	0·4352137608851296+01	−0·1145286894443553+01
$\gamma_{\text{unif}} - 1$	0·4210795187174558+01	−0·2000000000000226+01
0·001	0·3208603088724464+01	−0·1148603683480471+03
	$l = 5\lambda$	
10	0·1145337731736931+02	−0·7316319933534401−02
1	0·1098009635032070+02	−0·1112313610926676+00
0·1	0·1053691897894496+02	−0·9756791087418417+00
$\gamma_{\text{unif}} - 1$	0·1020641096155280+02	−0·4999999999998776+01
0·001	0·9546645868611446+01	−0·1109277148697554+03

carried out to sixteen significant figures for checking purposes). The value of γ_{unif} as computed from (8.7) is approximately 1·052 699 for $l = 2\lambda$ and 1·020 641 for $l = 5\lambda$. Eight terms were used in the series for $l = 2\lambda$ and eleven terms for $l = 5\lambda$, corresponding to $N = 14$ and 20, respectively. Only the even terms are present because the pattern is symmetric about the broadside direction $\eta = 0$.

In all of the cases shown the value of the arbitrary coefficient a_0 was chosen in such a way as to make the peak of the main beam have unit value. Thus the relative magnitudes of the current required to produce a signal of unit strength in the direction of maximum radiation for different values of γ_0 can be compared directly. For example, the peak current required on the 5λ aperture (at its ends) for $\gamma_0 - 1 = 10$ is over 23 times as great as for the uniform distribution. The maximum directivity obtainable, on the other hand, is only 1·122 times as great (approximately 0·5 dB).

The characteristic effects of superdirectivity are clearly evident in these figures: For large values of γ_0 the pattern becomes very large in the invisible region ($|\eta| > 1$) while changing but little in the visible region ($|\eta| < 1$).

And the aperture distribution fluctuates wildly over the aperture, rising to very high values at the ends.

Perhaps it should be emphasized that the pattern and aperture functions shown in Figs. 8.3 and 8.4 for the case $\gamma_0 = \gamma_{\text{unif}}$ were not obtained simply by plotting $(\sin u)/u$ and a constant, respectively, but were computed by exactly the same process as for all of the other values of γ_0. The fact that these particular results, obtained from the spheroidal expansions (8.11) and (8.34), do agree so precisely with those of the uniform distribution is further evidence of the validity of the optimum solution derived here and of the accuracy of the computations shown.

It is of interest to observe from Fig. 8.3 the way in which the pattern and aperture functions both approach their limiting functional form as γ_0 approaches its lower bound $\gamma_{00}(c)$. The limiting functional form of the pattern and aperture functions is shown in Appendix 3 to be proportional to the spheroidal functions $\psi_{00}(c, \eta)$ and $\psi_{00}(c, t)$, respectively. For the smallest of the five values of γ_0 used, which is close to the absolute lower bound $\gamma_{00}(2\pi) = 1{\cdot}000\,057\,250$ for $l = 2\lambda$, the pattern and aperture functions are each beginning to exhibit the general features of the limiting spheroidal function.

This solution of the maximum directivity problem settles an open question of long standing in antenna theory, in that it provides the first quantitative indication of the deteriorating effects of superdirectivity on the operating performance of actual antennas. From the numerical data shown for $\alpha = 0$ it is seen that the maximum directivity available in the broadside direction increases from its absolute lower bound at $\gamma_0 = \gamma_{00}(c)$ roughly as a logarithmic function of $\gamma_0 - 1$. This indicates that the rate at which maximum directivity increases with γ_0 must become very small for values of γ_0 greater than roughly $1{\cdot}1$, at which the value of maximum directivity is not much greater than the directivity of the uniform distribution. Thus, it is concluded that there is no hope of ever achieving directivities in practice that are appreciably greater than that of the uniform distribution, because to do so would require values of γ_0, and hence of the frequency sensitivity of the antenna, that are prohibitively great.

CHAPTER 9

MINIMUM PATTERN-SHAPING ERROR[1,12]

THE synthesis solutions obtained in the two preceding chapters apply only to the design of narrow-beam antennas. A different and much broader class of synthesis problems is that of shaped-beam design, any particular solution of which will depend upon the particular error criterion chosen. Although narrow-beam design does involve beam shaping of sorts, it differs from shaped-beam design in that the objective is not shape itself but rather some auxiliary property, usually resolution or directivity, that is associated with shape. In the case of true shaped-beam design the objective is to produce a pattern that provides the best approximation, in some specified sense, to a given pattern over the entire visible region. Usually the objective is to minimize some error measure for the radiation pattern as a whole. A physically interesting class of such solutions will be developed here.

For an aperture of any given length it is always possible, in principle, to find a source whose pattern space factor can be made to come as close as desired to any given function, whether aperture-limited or not, over the entire visible region. This was established by Bouwkamp and de Bruijn, and in a somewhat different way by Woodward and Lawson, as mentioned earlier. But in practice there is an absolute lower bound on the pattern error, or upper bound on pattern accuracy, for any prescribed value of the quality factor Q. This upper bound on pattern accuracy is produced by exactly the same physical phenomenon as that which produced the upper bound on directivity found in Chapter 8: the reactive power of the radiating source increases as the square of the magnitude of the invisible part of the pattern space factor, the latter of which becomes unbounded whenever the visible part is forced to represent a non-aperture-limited function. But the ratio of reactive to radiative power, and thence the value of Q, is limited physically by the input bandwidth required. Hence the prescribed value of Q will impose an absolute upper bound upon both the directivity and the pattern accuracy available from an aperture of any given length. The futility of any attempt to exceed the upper bound on directivity was established in Chapter 8. It will now be shown that any attempt to exceed the upper bound on pattern accuracy is equally futile.

9.1. Pattern error measure

Unlike directivity there is no unique measure for pattern error. Any quantity whose value depends upon the extent to which the desired pattern function differs from its aperture-limited approximation over the visible region could be used as a possible error measure. At present, however, there is only one such quantity for which an extremum has been found and for which a proof exists that the extremum is, in fact, the absolute lower bound on pattern error under the conditions specified. This is the mean-squared error M_α with weight factor $(1-\eta^2)^\alpha$, defined by

$$M_\alpha = \int_{-1}^{1} |F_\alpha(c\eta)-\hat{F}(c\eta)|^2 (1-\eta^2)^\alpha \, d\eta, \tag{9.1}$$

where $\hat{F}(c\eta)$ denotes the desired non-aperture-limited pattern function that is to be approximated by the aperture-limited pattern space factor $F_\alpha(c\eta)$. The source that minimizes this pattern error measure will be derived in the next section.

An important physical feature of this particular pattern error measure is the fact that it becomes proportional to the true radiated power that would be contained in the error pattern $F_\alpha(c\eta)-\hat{F}(c\eta)$ of E- or H-plane strip sources or of H-plane line sources when the value of α is chosen to be $-\frac{1}{2}$, $+\frac{1}{2}$, or 1, respectively. This is evident from (5.16), (5.20), and (5.29). These values of α are the same ones for which the directivity parameter D_α defined in the previous chapter was seen to become the true directivity, and for which the constraining parameter γ_α was either exactly or approximately related to the quality factor Q, for the case of strip and line sources. Hence the optimum pattern-shaping solution that will be obtained here is the natural companion to the maximum directivity solution that was obtained in the previous chapter.

The mathematical framework on which the optimum pattern-shaping solution will be based is very similar to that used in the previous chapter to obtain the maximum directivity solution. But there the similarity ends. The upper bound on directivity was found to depend solely upon the value assigned to three arbitrarily prescribed numbers, namely, the electrical length l/λ, the constraining parameter γ_α, and the exponent α. The lower bound on pattern error, on the other hand, must depend, in addition, upon an arbitrarily prescribed function, namely, the pattern function $\hat{F}(c\eta)$ to be approximated. Thus, whereas numerical solutions for the maximum directivity problem can be obtained once and for all, in principle, by a systematic program of computation over the range of all physically interesting values of l/λ, γ_α, and α, numerical solutions for the minimum pattern-error problem cannot. They must be recomputed for each individual pattern function of interest.

9.2. Optimum solution

An optimum solution to the pattern-shaping problem will now be obtained that achieves the absolute lower bound on the error measure M_α for any given value of l/λ, γ_α, and α. The objective is to find the source $A(t)$ whose pattern space factor $F_\alpha(c\eta)$ gives the best approximation, in the sense of minimum error M_α, to the given pattern function $\hat{F}(c\eta)$ over the visible region $-1 < \eta < 1$.

The pattern space factor required will be the aperture-limited function $F_\alpha(c\eta)$ that minimizes the constrained error functional

$$M_\alpha[F_\alpha] = \int_{-1}^{1} |F_\alpha(c\eta) - \hat{F}(c\eta)|^2 (1-\eta^2)^\alpha \, d\eta +$$
$$+\mu\left\{\int_{-\infty}^{\infty} |F_\alpha(c\eta)|^2 \, |1-\eta^2|^\alpha \, d\eta - \gamma_\alpha \int_{-1}^{1} |F_\alpha(c\eta)|^2 (1-\eta^2)^\alpha \, d\eta\right\}, \quad (9.2)$$

and that satisfies simultaneously the constraining condition (5.35). As in the formulation of the maximum directivity problem the parameter μ is a Lagrange multiplier whose value is to be determined.

By representing the unknown aperture space factor $f(t)$ as an expansion in an infinite series of spheroidal functions $\{\psi_{\alpha n}(c, t)\}$ of order α,

$$A(t) = (1-t^2)^\alpha f(t)$$
$$= (1-t^2)^\alpha \sum_{n=0}^{\infty} \frac{a_n}{\nu_{\alpha n}(c)} \psi_{\alpha n}(c, t), \qquad -1 \leqslant t \leqslant 1, \quad (9.3)$$

the resulting pattern space factor will also be represented exactly by an expansion in spheroidal functions,

$$F_\alpha(c\eta) = \sum_{n=0}^{\infty} a_n \psi_{\alpha n}(c, \eta), \qquad -\infty \leqslant \eta \leqslant \infty \quad (9.4)$$

by using the integral representation (5.11) for $\psi_{\alpha n}(c, \eta)$ after substituting (9.3) into (5.9) and integrating term by term. Furthermore the desired non-aperture-limited function $\hat{F}(c\eta)$ to be approximated by $F_\alpha(c\eta)$ can be expanded exactly in the same spheroidal functions over the visible region alone,

$$\hat{F}(c\eta) = \sum_{n=0}^{\infty} \hat{a}_n \psi_{\alpha n}(c, \eta), \qquad -1 < \eta < 1. \quad (9.5)$$

These expansion coefficients $\{\hat{a}_n\}$ can be obtained directly from the desired function by using the orthogonality property (5.12*a*),

$$\hat{a}_n = \frac{1}{\Lambda_{\alpha n}(c)} \int_{-1}^{1} \hat{F}(c\eta) \psi_{\alpha n}(c, \eta)(1-\eta^2)^\alpha \, d\eta, \quad (9.6)$$

provided only that $\hat{F}(c\eta)$ is square-integrable on $(-1, 1)$ with weight factor $(1-\eta^2)^\alpha$. Thus the optimization problem can be restated as being one of finding the set of coefficients $\{a_n\}$ that minimizes the constrained error functional obtained from (9.2),

$$M_\alpha[a_0, a_1, a_2, \ldots] = \sum_{n=0}^{\infty} [\{1+\mu(\gamma_{\alpha n}-\gamma_\alpha)\}a_n a_n^* - (a_n \hat{a}_n^* + a_n^* \hat{a}_n) + \hat{a}_n \hat{a}_n^*]\Lambda_{\alpha n}, \tag{9.7}$$

for an arbitrarily prescribed value of γ_α, and that satisfies simultaneously the constraining side condition (5.35). The crucial reduction from a double sum to a single sum in (9.7) occurred because of the double orthogonality relations (5.12) for the spheroidal functions.

A necessary condition for the constrained error (9.7) to have its smallest possible value is that it be an extremum with respect to each of the expansion coefficients $\{a_n\}$ separately. By differentiating it with respect to the complex conjugate of each coefficient and equating each derivative to zero the coefficients of the optimum solution are found to be simply

$$a_n = \frac{\hat{a}_n}{1+\mu(\gamma_{\alpha n}-\gamma_\alpha)}. \tag{9.8}$$

The Lagrange multiplier μ is a unique positive number less than $1/(\gamma_\alpha - \gamma_{\alpha 0})$ that satisfies the constraining side condition

$$h(\mu) = \sum_{n=0}^{\infty} \frac{|\hat{a}_n|^2 (\gamma_{\alpha n}-\gamma_\alpha)\Lambda_{\alpha n}}{\{1+\mu(\gamma_{\alpha n}-\gamma_\alpha)\}^2} = 0, \tag{9.9}$$

obtained from (5.35) by again using the series expansion (9.4) and the double orthogonality relations (5.12) for the spheroidal functions.

That this is, in fact, the solution that gives the absolute minimum value of M_α and not some other extremum can be established directly as follows. For any set of values $\{a_n+\delta_n\}$ of the expansion coefficients, each differing from the extremal value (9.8) by some complex number δ_n, the constrained error (9.7) becomes

$$M_\alpha = \sum_{n=0}^{\infty} \left[\frac{|\hat{a}_n|^2 \mu(\gamma_{\alpha n}-\gamma_\alpha)}{1+\mu(\gamma_{\alpha n}-\gamma_\alpha)} + |\delta_n|^2 \{1+\mu(\gamma_{\alpha n}-\gamma_\alpha)\} \right] \Lambda_{\alpha n}, \tag{9.10}$$

in which the value of μ still must satisfy the constraining side condition (9.9). The set of undetermined numbers $\{\delta_n\}$ must be such that the constraining parameter γ_α has its prescribed value, but is completely arbitrary otherwise. A necessary and sufficient condition for the extremum to be the absolute minimum sought is that the value of μ found from (9.9) be such that

$$1+\mu(\gamma_{\alpha n}-\gamma_\alpha) > 0, \qquad n = 0, 1, 2, 3, \ldots, \tag{9.11}$$

since the absolute minimum value of (9.10) would then occur when and only when all of the numbers $\{\delta_n\}$ are zero. And (9.11) will be satisfied if and only if

$$0 < \mu < 1/(\gamma_\alpha - \gamma_{\alpha 0}), \tag{9.12}$$

since $\gamma_{\alpha 0} < \gamma_{\alpha 1} < \gamma_{\alpha 2} < \dots \infty$. Thus, by requiring that the value of μ obtained from (9.9) lie within the interval (9.12) the extremum obtained is guaranteed to be the absolute minimum.

Condition (9.12) establishes, in addition, the truth of the assertion that the required solution of (9.9) will always be a unique positive number less than $1/(\gamma_\alpha - \gamma_{\alpha 0})$. The fact that it must be a positive number less than $1/(\gamma_\alpha - \gamma_{\alpha 0})$ is apparent. Its uniqueness follows directly from the observation that the function $h(\mu)$ decreases monotonically from the positive value $h = \infty$ that it has at $\mu = 0$ to the negative value $h = -\infty$ that it has at $\mu = 1/(\gamma_\alpha - \gamma_{\alpha 0})$, because of the fact that the slope of $h(\mu)$,

$$h'(\mu) = -2 \sum_{n=0}^{\infty} \frac{|\hat{a}_n|^2 (\gamma_{\alpha n} - \gamma_\alpha)^2 \Lambda_{\alpha n}}{\{1 + \mu(\gamma_{\alpha n} - \gamma_\alpha)\}^3}, \tag{9.13}$$

does not change sign for values of μ contained within the interval (9.12).

Thus, it is concluded that the series expansion (9.4) with coefficients (9.8) gives the best mean-square approximation with weight factor $(1-\eta^2)^\alpha$ to any given function $\hat{F}(c\eta)$ that is possible to achieve for an arbitrarily prescribed value of the parameters c, γ_α, and α.

The constraining condition (9.9) can be solved numerically for the Lagrange multiplier μ by noting that, to the first order, it can be written as

$$h(\mu) + h'(\mu)\,\Delta\mu = 0. \tag{9.14}$$

Hence, for any approximate value μ_a lying within the required interval (9.12) the first-order correction $\Delta\mu_a$ will be simply

$$\Delta\mu_a = -h(\mu_a)/h'(\mu_a). \tag{9.15}$$

The sequence of approximate values μ_a obtained from these first-order corrections has been found to converge readily to the required solution of (9.9).

The above solution of the optimum pattern-shaping problem was based on the assumption that the radiation pattern desired has been specified in phase as well as in magnitude. Indeed, there are important practical cases in which phase must be specified, such as in the design of reflector feeds. But more often than not the magnitude alone is all that is specified, phase being of no particular interest. This would be the case, for example, in the design of sources for search radar applications. In such cases it is almost invariably assumed (usually implicitly) that the phase pattern is constant. There is no assurance, however, that a constant phase-pattern is anywhere near the best choice. To obtain the true optimum solution in such cases would require that the phase of the given radiation pattern be chosen, instead, in such a way as to reduce the error measure M_α obtained above to its lowest possible value for the particular pattern magnitude function specified. A necessary condition on the phase $\phi(c\eta)$ of the given pattern function $\hat{F}(c\eta)$ can be obtained by varying $\phi(c\eta)$ by an amount $\epsilon\zeta(c\eta)$ and insisting that as ϵ approaches zero the derivative of the error measure M_α with respect to ϵ must vanish for an arbitrarily chosen function $\zeta(c\eta)$. By this process the following condition on the phase pattern $\phi(c\eta)$ for any prescribed pattern-magnitude $|\hat{F}(c\eta)|$ can be obtained,

$$\int_{-1}^{1} |\hat{F}(c\eta)| \left[\sum_{n=0}^{\infty} \frac{(\gamma_{\alpha n}-\gamma_\alpha)\psi_{\alpha n}(c,\eta)\psi_{\alpha n}(c,\eta')}{\{1+\mu(\gamma_{\alpha n}-\gamma_\alpha)\}\Lambda_{\alpha n}}\right] \sin\{\phi(c\eta)-\phi(c\eta')\}(1-\eta^2)^\alpha\, d\eta \underset{\eta'}{\equiv} 0, \tag{9.16}$$

for η' lying anywhere within the interval $(-1, 1)$. It is evident that this condition would be satisfied by choosing any constant value for the phase function $\phi(c\eta)$. But there is no assurance that a constant phase function will produce the smallest error possible for a given pattern-magnitude function since (9.16) is only a necessary condition. Thus, the problem of choosing the best phase function to go with a given pattern-magnitude function still appears to remain unsolved.

9.3. Illustration: cosecant-pattern approximation for *E*-plane strip sources

The optimum solution above will now be illustrated numerically for the case of an *E*-plane strip source of length a that is designed to produce the best *Q*-constrained approximation to the cosecant pattern function $\hat{F}_{xx}(c_x n_x)$ defined earlier by (6.5). Since the element factor is unity in the *E*-plane this represents a true cosecant radiation pattern. And since an exact relationship exists between the constraining parameter γ_α for $\alpha = -\frac{1}{2}$ and the value of Q for an *E*-plane strip source the resulting solution is a true *Q*-constrained solution.

The spheroidal functions that lead to a *Q*-constrained approximation for *E*-plane strip sources are those of order $\alpha = -\frac{1}{2}$. They are exactly the same as the even Mathieu functions $\{Se_n(c_x, n_x)\}$ that were proposed by Pistolkors† and Leonard.‡ But it should be noted that the synthesis solution obtained by Pistolkors and Leonard gave just the unconstrained source that would be obtained by using the expansion coefficients (9.6) of the non-aperture-limited cosecant radiation pattern $\hat{F}_{xx}(c_x n_x)$. The lack of a constraint causes such a solution to become unrealizable physically when the number of terms

† A. A. Pistolkors, *loc. cit.* ‡ D. J. Leonard, *loc. cit.*

in the series is allowed to become appreciably greater than twice the number of wavelengths contained within the length of the aperture. This is illustrated by the pattern approximation to the cosecant function shown in Fig. 9.1 for an aperture of length $a = 2{\cdot}5\ \lambda$. It was computed for a 35-term series of spheroidal functions of order $\alpha = -\frac{1}{2}$ whose coefficients $\{a_n\}$ were simply the unconstrained coefficients $\{\hat{a}_n\}$ obtained from (9.6). Over the visible region shown it provides an excellent approximation to the cosecant function desired. But outside of the visible region its analytic continuation was found to rise to enormous values that exceeded 10^{20}. Thus the source required to produce such a pattern approximation would be overwhelmingly reactive and, hence, totally unrealizable physically. This is another vivid illustration of the physical limitations that are inherent in all unconstrained synthesis solutions.

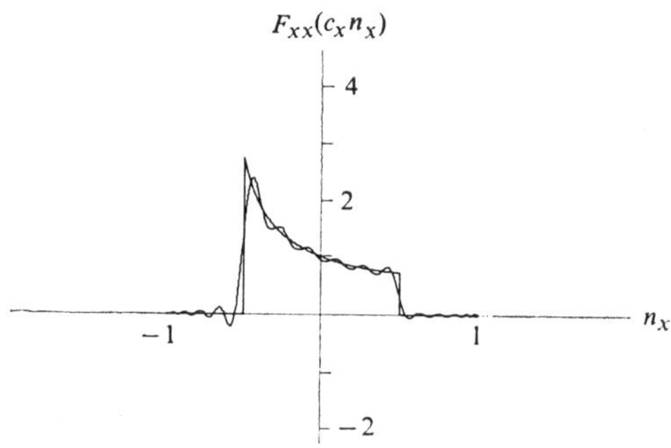

FIG. 9.1. Unconstrained 35-term spheroidal function approximation to the cosecant pattern (6.5) for an E-plane strip source of length $a = 2{\cdot}5\lambda$ for $\alpha = -\frac{1}{2}$.

If the value of Q used to obtain the constrained synthesis solution in section 9.2 were allowed to become arbitrarily large the amount of reactive power would again be unconstrained and the resulting solution would be like that in Fig. 9.1. But if the amount of reactive power were constrained by restricting the value of Q to be 1000, 10, or 0·1 the optimum pattern approximation and its aperture distribution for the same aperture length $a = 2{\cdot}5\lambda$ used in Fig. 9.1 would become as shown in Fig. 9.2. The magnitude of the reactive part of the pattern space factor (the invisible region, shown broken) decreases with decreasing values of Q, as does also the magnitude of the aperture distribution, but only at the cost of reduced precision of the pattern approximation over the visible region (the solid line). These plots can be expected to be a reasonably accurate portrayal of the kind of physical results actually attainable in practice, except in the immediate neighbourhood of the edges where the solution necessarily becomes non-physical because of a non-physical assumption on the value assigned to the edge exponent α. It is evident from these results that relatively little improvement in the pattern approximation can be expected by going to a high-Q design.

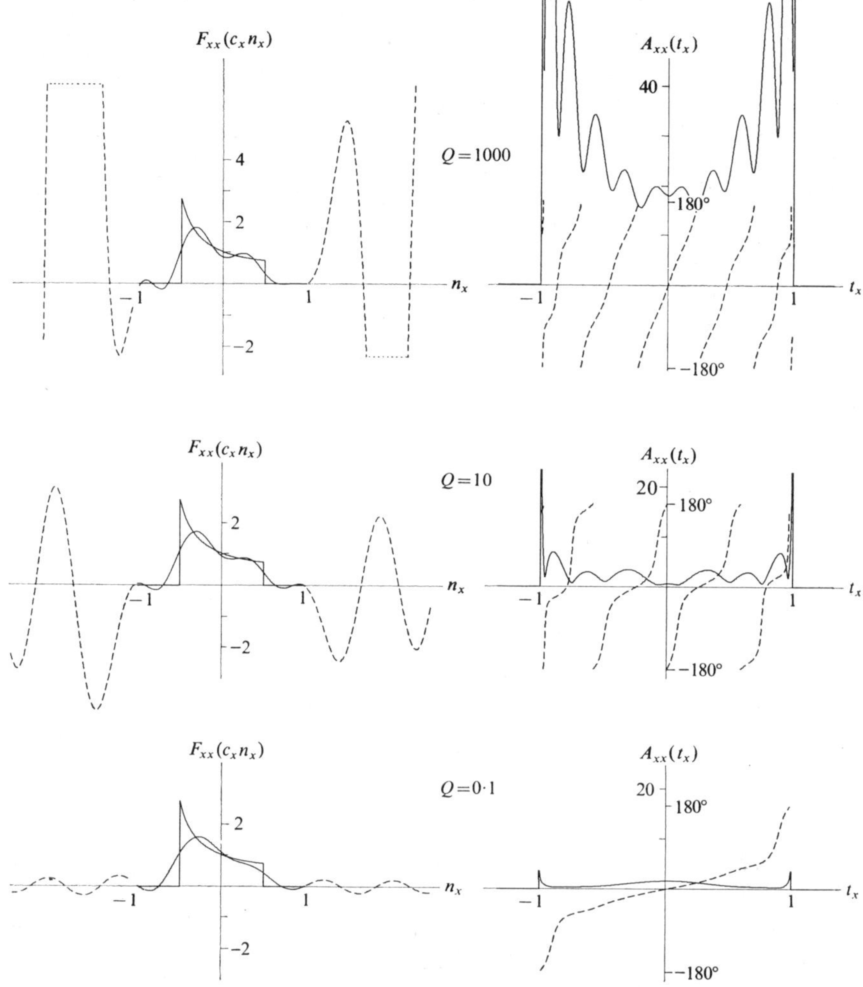

FIG. 9.2. The optimum Q-constrained approximation (9.4) to the cosecant radiation pattern (6.5), and the magnitude (solid) and phase (broken) of its aperture distribution (9.3), for an E-plane strip source of length $a = 2{\cdot}5\lambda$, with $\alpha = -\frac{1}{2}$ and $Q = \gamma_{-\frac{1}{2}}-1 = 1000$, 10, and 0·1.

The extent to which the pattern approximation can be improved by increasing the length of the aperture, however, is quite a different story. Pattern accuracy increases rapidly with aperture length. This is illustrated by the syntheses shown in Fig. 9.3 for three different aperture lengths $a = 5\lambda$, $2{\cdot}5\lambda$, and λ when the value of Q is held fixed at $Q = 10$. The pattern approximation is clearly far more accurate for an aperture of length 5λ than for an aperture of length λ.

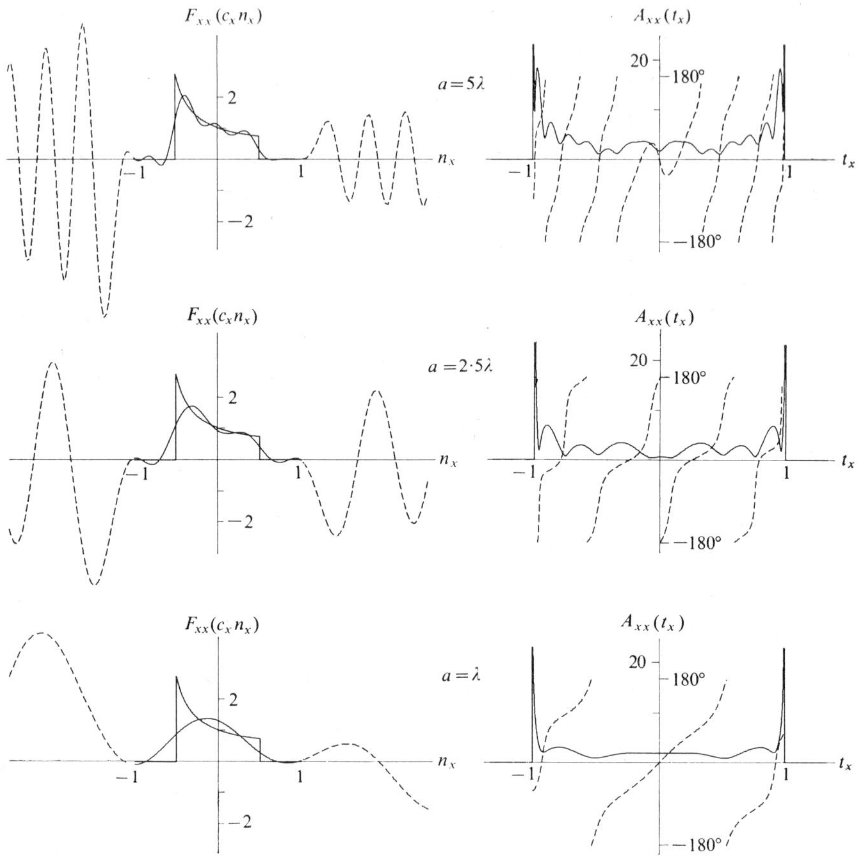

FIG. 9.3. The optimum Q-constrained approximation (9.4) to the cosecant radiation pattern (6.5), and the magnitude (solid) and phase (broken) of its aperture distribution (9.3), for an E-plane strip source of length $a = 5\lambda$, $2\cdot5\lambda$, and λ, with $\alpha = -\frac{1}{2}$ and $Q = \gamma_{-\frac{1}{2}}-1 = 10$.

Although the accuracy of the pattern approximation can be increased by increasing either the value of Q or of the aperture length a/λ it is worthwhile noting from Fig. 9.3 that the deleterious features appearing in Fig. 9.2 for a high-Q design do not appear for a long-aperture design. A large increase in the strength of the source is required for a high-Q design, whereas for a long-aperture design the strength of the source remains almost unchanged. Also, a substantial increase in the rapidity with which the magnitude and phase of the source must fluctuate over its length is required for a high-Q design, whereas the rapidity of the magnitude and phase fluctuations required for a long-aperture design remains almost unchanged (remember that the source is shown here as a function of the normalized variables $t_x = 2x/a$, which means that for $a = 5\lambda$ it must be stretched out over an interval five times as long as for $a = \lambda$).

Perhaps it should be emphasized once more that the synthesis solution on which these calculations were based provides an absolute upper bound on the pattern accuracy available. Thus, the pattern approximations shown in Figs. 9.2 and 9.3 are the best that are possible to achieve for the assigned values of electrical length a/λ and of quality factor Q of an E-plane strip source, in the sense of absolute minimum power that would be radiated by the error pattern.

9.4. Conclusion

By a fortuitous combination of circumstances the parameter γ_α has come to play a remarkable role in the theory of strip and line source synthesis as developed in this treatise. First came the discovery of some purely mathematical properties, as outlined in Appendix 3, that wove it tightly into the theory of the spheroidal functions $\{\psi_{\alpha n}(c, \eta)\}$ of order α: It had precisely the features needed to take full advantage of the double orthogonality property (5.12). It was intimately related to a third kind of characteristic numbers $\{\gamma_{\alpha n}(c)\}$ that arose first from the double orthogonality of the spheroidal functions and again from the characteristic numbers of a new integral equation for the spheroidal functions, a relationship that led, incidentally, to the particular choice of notation for those numbers. And it was found to possess an absolute lower bound $\gamma_{\alpha 0}(c)$ arising from an important extremal property of the spheroidal functions. Next came certain physical properties, developed in section 5.5, that wove it into the body of theory associated with the quality factor Q of strip and line sources: It was found to be related exactly to the Q of E-plane strip sources by choosing the value of α to be $-\frac{1}{2}$, and to be related at least approximately to the Q of H-plane strip and line sources by choosing the value of α to be $\frac{1}{2}$ and 1, respectively. Then it led to the physical solution obtained in Chapter 8 for the maximum directivity available from strip and line sources when constrained to have an arbitrarily prescribed value of the quality factor Q. And it has now led to a physical solution to the optimum pattern-shaping problem for strip and line sources in the sense of least radiated power in the error pattern when constrained again to have an arbitrarily prescribed value of the quality factor Q. Thus, the parameter γ_α has made possible a thoroughly physical solution, except for edge behaviour, to some of the most important problems of strip and line source synthesis. And it has done so in a form for which precise numerical results are readily obtainable.

APPENDIX 1

ASYMPTOTIC BEHAVIOUR OF THE FIELD FUNCTIONS

A common criticism of the method of stationary phase for determining the asymptotic behaviour of integrals of the form (2.9) is that it is not very convincing. The purpose of this Appendix is to suggest that it might be made more convincing by means of an asymptotic evaluation based upon the Riemann–Lebesque lemma, as follows.

The integral (2.9) will be written as

$$\Phi(x, y, z, k) = \frac{1}{(2\pi)^2} \int_{-\infty}^{\infty} \int_{-\infty}^{\infty} F(k_x, k_y, k) e^{-i(k_x x + k_y y + k_z z)}\, dk_x\, dk_y, \qquad \text{(A1.1)}$$

in which

$$k_z = \begin{cases} (k^2 - k_x^{\,2} - k_y^{\,2})^{\frac{1}{2}}, & k_x^{\,2} + k_y^{\,2} < k^2 \\ -i(k_x^{\,2} + k_y^{\,2} - k^2)^{\frac{1}{2}}, & k_x^{\,2} + k_y^{\,2} > k^2. \end{cases} \qquad \text{(A1.2)}$$

The objective is to determine its asymptotic value in any given direction (θ, ϕ) of Fig. 2.1 as the distance $r = (x^2+y^2+z^2)^{\frac{1}{2}}$ approaches infinity.

Changing from rectangular coordinates (x, y, z) to polar coordinates (r, θ, ϕ) and from (k_x, k_y) to (α, β) the integral (A1.1) becomes

$$\Phi(r \sin\theta \cos\phi, r \sin\theta \sin\phi, r\cos\theta, k)$$

$$= \frac{1}{(2\pi)^2} \int_{C_\alpha} \int_{C_\beta} F(k \sin\alpha \cos\beta, k \sin\alpha \sin\beta, k) \times$$

$$\times \exp\{-ikr(\sin\alpha \cos\beta \sin\theta \cos\phi +$$

$$+ \sin\alpha \sin\beta \sin\theta \sin\phi + \cos\alpha \cos\theta)\} \times$$

$$\times k^2 \sin\alpha \cos\alpha \, d\alpha\, d\beta, \qquad \text{(A1.3)}$$

where C_α and C_β denote the α- and β-plane contours, respectively, shown in Fig. 2.5c. The total contribution from the visible region $-1 < \sin\alpha < 1$ will be seen shortly to decay asymptotically as $1/r$. But in the invisible region $|\sin\alpha| > 1$, where $\cos\alpha = -i(\sin^2\alpha - 1)^{\frac{1}{2}}$, the factor

$$\exp\{-ikr \cos\alpha \cos\theta\} = \exp\{-kr(\sin^2\alpha - 1)^{\frac{1}{2}} \cos\theta\} \qquad \text{(A1.4)}$$

causes each infinitesimal contribution from the integrand to vanish exponentially with increasing r. Hence the total contribution from the invisible region vanishes faster than any power of r.

As r approaches infinity, then, the integral (A1.3) becomes an integral over just the visible region alone, which can be rewritten as

$$\Phi \sim \frac{k^2 e^{-ikr}}{(2\pi)^2} \int\limits_{-\frac{\pi}{2}}^{\frac{\pi}{2}} \left(\int\limits_{-\frac{\pi}{2}}^{\frac{\pi}{2}} F \exp[ikr\{1-\cos(\beta-\phi)\}\sin\alpha\sin\theta]\, d\beta \right) \times$$

$$\times \exp[ikr\{1-\cos(\alpha-\theta)\}]\sin\alpha\cos\alpha\, d\alpha. \quad \text{(A1.5)}$$

The inner integral will be evaluated first.† In order to cast it into a form to which the Riemann–Lebesque lemma can be applied the variable β will be changed to $\xi = k\{1-\cos(\beta-\phi)\}\sin\alpha\sin\theta$,

$$\int\limits_{-\frac{\pi}{2}}^{\frac{\pi}{2}} F \exp[ikr\{1-\cos(\beta-\phi)\}\sin\alpha\sin\theta]\, d\beta$$

$$= \frac{1}{k\sin\alpha\sin\theta} \left(\int\limits_{k(1+\sin\phi)\sin\alpha\sin\theta}^{0} \frac{F}{\{1-(1-\xi/k\sin\alpha\sin\theta)^2\}^{\frac{1}{2}}} e^{ir\xi}\, d\xi + \right.$$

$$\left. + \int\limits_{0}^{k(1-\sin\phi)\sin\alpha\sin\theta} \frac{F}{\{1-(1-\xi/k\sin\alpha\sin\theta)^2\}^{\frac{1}{2}}} e^{ir\xi}\, d\xi \right), \quad \text{(A1.6)}$$

the first interval corresponding to $-\frac{1}{2}\pi < \beta < \phi$ and the second to $\phi < \beta < \frac{1}{2}\pi$. The Riemann–Lebesque lemma asserts‡ that if any function $f(\xi)$ has limited total fluctuation on the interval (a, b) then, as $r \to \infty$,

$$\int\limits_{a}^{b} f(\xi) e^{ir\xi}\, d\xi = O(1/r). \quad \text{(A1.7)}$$

But the function $F/\{1-(1-\xi/k\sin\alpha\sin\theta)^2\}^{\frac{1}{2}}$ in each of the two integrals in (A1.6) is not bounded at the endpoint $\xi = 0$, so the lemma does not apply directly. Instead, remove that endpoint from each of the integrals by removing the part of the integral that comes from the small interval $\phi-\epsilon_\beta < \beta < \phi+\epsilon_\beta$

† It will be assumed here that $\theta \neq 0$, for purposes of evaluating the integrals in (A1.5).
‡ E. T. Whittaker and G. N. Watson, *loc. cit.*, p. 172.

about the point $\beta = \phi$ (the 'point of stationary phase'), where ϵ_β is an arbitrarily small positive number:

$$\int_{-\frac{\pi}{2}}^{\frac{\pi}{2}} F \exp[ikr\{1-\cos(\beta-\phi)\}\sin\alpha\sin\theta]\,d\beta$$

$$= \int_{\phi-\epsilon_\beta}^{\phi+\epsilon_\beta} F \exp[ikr\{1-\cos(\beta-\phi)\}\sin\alpha\sin\theta]\,d\beta +$$

$$+\frac{1}{k\sin\alpha\sin\theta}\left(\int_{k(1+\sin\phi)\sin\alpha\sin\theta}^{k(1-\cos\epsilon_\beta)\sin\alpha\sin\theta} \frac{F}{\{1-(1-\xi/k\sin\alpha\sin\theta)^2\}^{\frac{1}{2}}} e^{ir\xi}\,d\xi + \right.$$

$$\left. + \int_{k(1-\cos\epsilon_\beta)\sin\alpha\sin\theta}^{k(1-\sin\phi)\sin\alpha\sin\theta} \frac{F}{\{1-(1-\xi/k\sin\alpha\sin\theta)^2\}^{\frac{1}{2}}} e^{ir\xi}\,d\xi\right). \qquad \text{(A1.8)}$$

The function in the integrand of the latter two integrals does, indeed, have limited total fluctuation over both intervals (the F function itself always has limited total fluctuation since it is an entire analytic function, as shown in section 3.1), hence the latter two integrals are $O(1/r)$ functions. And by choosing ϵ_β sufficiently small the first integral can be approximated with arbitrarily high precision by

$$\int_{\phi-\epsilon_\beta}^{\phi+\epsilon_\beta} F \exp[ikr\{1-\cos(\beta-\phi)\}\sin\alpha\sin\theta]\,d\beta$$

$$\doteqdot F(k\sin\alpha\cos\beta,\, k\sin\alpha\sin\beta,\, k)\Big|_{\beta=\phi} \cdot \int_{\phi-\epsilon_\beta}^{\phi+\epsilon_\beta} \exp\left\{ikr\frac{(\beta-\phi)^2}{2}\sin\alpha\sin\theta\right\}d\beta, \qquad \text{(A1.9)}$$

the latter of which can be put into a standard form by a simple change of variable,

$$\int_{-\epsilon}^{\phi+\epsilon_\beta} \exp\left\{ikr\frac{(\beta-\phi)^2}{2}\sin\alpha\sin\theta\right\}d\beta$$

$$= 2\left(\frac{\pi}{kr\sin\alpha\sin\theta}\right)^{\frac{1}{2}} \int_0^{\epsilon_\beta\{(kr\sin\alpha\sin\theta)/\pi\}^{\frac{1}{2}}} \exp(i\tfrac{1}{2}\pi t^2)\,dt$$

$$= \left(\frac{2\pi}{kr\sin\alpha\sin\theta}\right)^{\frac{1}{2}} e^{i\frac{\pi}{4}}\left\{1+O\left(\frac{1}{\epsilon_\beta r^{\frac{1}{2}}}\right)\right\}. \qquad \text{(A1.10)}$$

By letting ϵ_β approach zero in such a way that $\epsilon_\beta r^{\frac{1}{2}}$ still approaches infinity the inner integral in (A1.5) then becomes

$$\int_{-\frac{\pi}{2}}^{\frac{\pi}{2}} F(k\sin\alpha\cos\beta, k\sin\alpha\sin\beta, k)\exp[ikr\{1-\cos(\beta-\phi)\}\sin\alpha\sin\theta]\,d\beta$$

$$= \left(\frac{2\pi}{kr\sin\alpha\sin\theta}\right)^{\frac{1}{2}} e^{i\frac{\pi}{4}} F(k\sin\alpha\cos\phi, k\sin\alpha\sin\phi, k)\left\{1+O\left(\frac{1}{r}\right)\right\}. \quad \text{(A1.11)}$$

The outer integral in (A1.5) can be evaluated in the same manner by first removing the part of the integral that comes from an arbitrarily small interval $\theta-\epsilon_\alpha < \alpha < \theta+\epsilon_\alpha$ about the point of stationary phase $\alpha = \theta$ and then applying the Riemann–Lebesque lemma to the rest. Thus, the asymptotic behaviour of the scalar function (A1.1) is found to be

$$\Phi(r\sin\theta\cos\phi, r\sin\theta\sin\phi, r\cos\theta, k)$$
$$\sim \frac{ik}{2\pi}\frac{e^{-ikr}}{r}\cos\theta\, F(k\sin\theta\cos\phi, k\sin\theta\sin\phi, k). \quad \text{(A1.12)}$$

This appears to provide a more rigorous basis for the principle of stationary phase.

APPENDIX 2

INTEGRALS ASSOCIATED WITH PLANAR DIPOLE ANTENNAS[2,4]

THE integrals on k_x in (3.11) can be written in the following form

$$\int_{(k^2-k_y^2)^{\frac{1}{2}}}^{\infty}\left(\frac{\sin\frac{1}{2}k_x a}{\frac{1}{2}k_x a}\right)^2\frac{dk_x}{\{k_x^2-(k^2-k_y^2)\}^{\frac{1}{2}}}$$

$$=\frac{4}{(k^2-k_y^2)a^2}\int_0^{\infty}\frac{\sin^2\{\frac{1}{2}a(k^2-k_y^2)^{\frac{1}{2}}\cosh\xi\}}{\cosh^2\xi}\,d\xi, \quad \text{(A2.1)}$$

$$\int_0^{\infty}\left(\frac{\sin\frac{1}{2}k_x a}{\frac{1}{2}k_x a}\right)^2\frac{dk_x}{\{k_x^2+(k_y^2-k^2)\}^{\frac{1}{2}}}$$

$$=\frac{4}{(k_y^2-k^2)a^2}\int_0^{\infty}\frac{\sin^2\{\frac{1}{2}a(k_y^2-k^2)^{\frac{1}{2}}\sinh\xi\}}{\sinh^2\xi}\,d\xi, \quad \text{(A2.2)}$$

by a change of variable $k_x = (k^2-k_y^2)^{\frac{1}{2}}\cosh\xi$ in the first integral and $k_x = (k_y^2-k^2)^{\frac{1}{2}}\sinh\xi$ in the second. Differentiating the first integral on ξ twice with respect to the parameter $a(k^2-k_y^2)^{\frac{1}{2}}$ and the second twice with respect to $a(k_y^2-k^2)^{\frac{1}{2}}$ one obtains, respectively,

$$\tfrac{1}{2}\int_0^{\infty}\cos\{a(k^2-k_y^2)^{\frac{1}{2}}\cosh\xi\}\,d\xi,$$

$$\tfrac{1}{2}\int_0^{\infty}\cos\{a(k_y^2-k^2)^{\frac{1}{2}}\sinh\xi\}\,d\xi,$$

the first of which is equal to $-\frac{1}{4}\pi Y_0\{a(k^2-k_y^2)^{\frac{1}{2}}\}$ and the second to $\frac{1}{2}K_0\{a(k_y^2-k^2)^{\frac{1}{2}}\}$. Integrating each of these twice to regain the original integral and then integrating by parts, the integrals (A2.1) and (A2.2) for a fixed value of k_y become

$$\int\limits_{(k^2-k_y^2)^{\frac{1}{2}}}^{\infty} \left(\frac{\sin \frac{1}{2}k_x a}{\frac{1}{2}k_x a}\right)^2 \frac{dk_x}{\{k_x^2-(k^2-k_y^2)\}^{\frac{1}{2}}}$$

$$= -\frac{\pi}{(k^2-k_y^2)a^2}\left[a(k^2-k_y^2)^{\frac{1}{2}} \int\limits_0^{a(k^2-k_y^2)^{\frac{1}{2}}} Y_0(x)\,dx - \right.$$

$$\left. -a(k^2-k_y^2)^{\frac{1}{2}} Y_1\{a(k^2-k_y^2)^{\frac{1}{2}}\} - \frac{2}{\pi}\right]$$

$$= [\tfrac{3}{2}-C-\ln\{\tfrac{1}{2}a(k^2-k_y^2)^{\frac{1}{2}}\}][1+O(a^2\ln a)\}, \qquad \text{(A2.3)}$$

$$\int\limits_0^{\infty} \left(\frac{\sin \frac{1}{2}k_x a}{\frac{1}{2}k_x a}\right)^2 \frac{dk_x}{\{k_x^2+(k_y^2-k^2)\}^{\frac{1}{2}}}$$

$$= \frac{2}{(k_y^2-k^2)a^2}\left[a(k_y^2-k^2)^{\frac{1}{2}} \int\limits_0^{a(k_y^2-k^2)^{\frac{1}{2}}} K_0(x)\,dx + \right.$$

$$\left. +a(k_y^2-k^2)^{\frac{1}{2}} K_1\{a(k_y^2-k^2)^{\frac{1}{2}}\} - 1\right]$$

$$= [\tfrac{3}{2}-C-\ln\{\tfrac{1}{2}a(k_y^2-k^2)^{\frac{1}{2}}\}][1+O(a^2\ln a)\}, \qquad \text{(A2.4)}$$

where C denotes Euler's constant 0·57721... .

The following integrals have been evaluated exactly in closed form in terms of sine and cosine integrals:

$$\int\limits_0^1 \frac{(\cos \frac{1}{2}kb\eta - \cos \frac{1}{2}kb)^2}{1-\eta^2}\,d\eta = \tfrac{1}{2}\{\operatorname{Cin} kb +$$

$$+(\operatorname{Cin} kb - \tfrac{1}{2}\operatorname{Cin} 2kb)\cos kb - (\operatorname{Si} kb - \tfrac{1}{2}\operatorname{Si} 2kb)\sin kb\}, \qquad \text{(A2.5)}$$

$$\int\limits_0^{\infty} \frac{(\cos \frac{1}{2}kb\eta - \cos \frac{1}{2}kb)^2}{1-\eta^2}\,d\eta = -\tfrac{1}{4}\pi \sin kb, \qquad \text{(A2.6)}$$

$$\int_0^\infty \frac{(\cos\frac{1}{2}kb\eta - \cos\frac{1}{2}kb)^2}{1-\eta^2} \ln|1-\eta^2|^{\frac{1}{2}}\,d\eta$$

$$= -\tfrac{1}{4}\pi[\mathrm{Si}\,kb + (\mathrm{Si}\,kb - \tfrac{1}{2}\,\mathrm{Si}\,2kb)\cos kb + {} $$
$$+ (\mathrm{Cin}\,kb - \tfrac{1}{2}\,\mathrm{Cin}\,2kb - C - \ln\tfrac{1}{4}kb)\sin kb], \quad \text{(A2.7)}$$

$$\int_0^\infty \left(\frac{\cos\frac{1}{2}kb\eta - \cos\frac{1}{2}kb}{1-\eta^2}\right)^2 d\eta = \tfrac{1}{8}\pi(kb - \sin kb), \quad \text{(A2.8)}$$

$$\int_0^\infty \left(\frac{\cos\frac{1}{2}kb\eta - \cos\frac{1}{2}kb}{1-\eta^2}\right)^2 \ln|1-\eta^2|^{\frac{1}{2}}\,d\eta$$

$$= -\tfrac{1}{8}\pi[\mathrm{Si}\,kb - \tfrac{1}{2}kb\,\mathrm{Cin}\,kb + (kb - \sin kb)(C - \tfrac{1}{2} + \ln\tfrac{1}{2}kb) + {}$$
$$+ \{(\mathrm{Si}\,kb - \tfrac{1}{2}\,\mathrm{Si}\,2kb) - \tfrac{1}{2}kb(\mathrm{Cin}\,kb - \mathrm{Cin}\,2kb) - kb\ln 2\}\cos kb + {}$$
$$+ \{(\mathrm{Cin}\,kb - \tfrac{1}{2}\,\mathrm{Cin}\,2kb) + \tfrac{1}{2}kb(\mathrm{Si}\,kb - \mathrm{Si}\,2kb) + \ln(2/e)\}\sin kb],$$
$$\text{(A2.9)}$$

where the functions Si x and Cin x are defined by

$$\mathrm{Si}\,x = \int_0^x \frac{\sin t}{t}\,dt, \quad \text{(A2.10)}$$

$$\mathrm{Cin}\,x = \int_0^x \frac{1-\cos t}{t}\,dt. \quad \text{(A2.11)}$$

APPENDIX 3
SPHEROIDAL FUNCTIONS[3,5,6]

ONE of two independent solutions of the spheroidal differential equation investigated by Stratton,†

$$(1-\eta^2)\psi_{\alpha n}''(c,\eta)-2(\alpha+1)\eta\psi_{\alpha n}'(c,\eta)+(b_{\alpha n}-c^2\eta^2)\psi_{\alpha n}(c,\eta)=0, \quad \text{(A3.1)}$$

is regular at the singular points $\eta = \pm 1$ if the parameter $b_{\alpha n}(c)$ is restricted to one of a discrete set of characteristic numbers, indexed by $n = 0, 1, 2, 3, \ldots,$ that depend upon the values assigned to the real parameters α and c. When this solution is sought in the form of the integral

$$\psi_{\alpha n}(c,\eta) = \int_{t_1}^{t_2} e^{ic\eta t}(1-t^2)^{\alpha}u(c,t)\,dt \qquad \text{(A3.2)}$$

it is found, upon introducing (A3.2) into (A3.1) and integrating by parts, that the limits of integration and the function $u(c, t)$ must be chosen to satisfy

$$e^{ic\eta t}(1-t^2)^{\alpha+1}(u'-ic\eta u)\Big|_{t=t_1}^{t_2} - \int_{t_1}^{t_2} e^{ic\eta t}(1-t^2)^{\alpha}\{(1-t^2)u''-2(\alpha+1)tu'+(b_{\alpha n}-c^2t^2)u\}\,dt \underset{\eta}{\equiv} 0. \qquad \text{(A3.3)}$$

The bracketed expression within the integral is of precisely the same form as the left-hand side of (A3.1), hence the integral will vanish identically in η if $u(c, t)$ is chosen to be proportional to $\psi_{\alpha n}(c, t)$. The first term will also vanish identically in η if the limits of integration are chosen to be ± 1, providing that $\alpha > -1$. Consequently the solutions of (A3.1) that are regular at its singular points are the solutions of the integral equation

$$\nu_{\alpha n}(c)\psi_{\alpha n}(c,\eta) = \int_{-1}^{1} e^{ic\eta t}(1-t^2)^{\alpha}\psi_{\alpha n}(c,t)\,dt \qquad \text{(A3.4)}$$

for any $\alpha > -1$, where the proportionality constant $\nu_{\alpha n}(c)$ is the characteristic number of (A3.4) that corresponds to the nth characteristic number $b_{\alpha n}(c)$ of the differential equation (A3.1).

From the differential equation (A3.1) and the boundary condition of finiteness at $\eta = \pm 1$ it follows that the spheroidal functions are entire functions of η, that they have no zeros at $\eta = \pm 1$, that they are real for real η,

† J. A. Stratton (1935); *loc. cit.*

that they have exactly n zeros within the interval $(-1, 1)$, and that they are even or odd functions of η according as n is even or odd. As the parameter c approaches zero the characteristic numbers $\{b_{\alpha n}(c)\}$ approach the value

$$b_{\alpha n}(c) \to n(n+2\alpha+1) \tag{A3.5}$$

and the spheroidal differential equation approaches the Gegenbauer differential equation,

$$(1-\eta^2)T_n^{\alpha\prime\prime}(\eta)-2(\alpha+1)\eta T_n^{\alpha\prime}(\eta)+n(n+2\alpha+1)T_n^{\alpha}(\eta) = 0. \tag{A3.6}$$

Stratton's form of the Gegenbauer polynomials $\{T_n^{\alpha}(\eta)\}$ for $\alpha = m = 0, 1, 2, 3, \ldots$ is related to the associated Legendre functions $\{P_{m+n}^m(\eta)\}$ by

$$T_n^m(\eta) = \frac{P_{m+n}^m(\eta)}{(1-\eta^2)^{m/2}}, \tag{A3.7}$$

and for $\alpha = -\frac{1}{2}$ or $+\frac{1}{2}$ it is related to the Chebyshev polynomials of the first and second kinds, respectively, by

$$nT_n^{-\frac{1}{2}}(\eta) = (2/\pi)^{\frac{1}{2}} \cos\{n \operatorname{arc} \cos \eta\}, \tag{A3.8}$$

$$T_n^{\frac{1}{2}}(\eta) = (2/\pi)^{\frac{1}{2}} \frac{\sin\{(n+1)\operatorname{arc} \cos \eta\}}{(1-\eta^2)^{\frac{1}{2}}}. \tag{A3.9}$$

Also the characteristic numbers $\{b_{\alpha n}(c)\}$ are real and ordered such that

$$b_{\alpha 0} < b_{\alpha 1} < b_{\alpha 2} < b_{\alpha 3} < \ldots . \tag{A3.10}$$

From the integral representation (A3.4) for the spheroidal functions it follows that they are aperture-limited functions of order α and that their characteristic numbers $\{\nu_{\alpha n}(c)\}$ are real for even n and imaginary for odd n,

$$\nu_{\alpha n}(c) = i^n\,|\nu_{\alpha n}(c)|. \tag{A3.11}$$

And from the asymptotic behaviour (6.30) possessed by such integrals it is evident that the asymptotic behaviour of $\psi_{\alpha n}(c, \eta)$ as η approaches infinity must be

$$\psi_{\alpha n}(c, \eta) = \frac{2^{\alpha+1}\Gamma(\alpha+1)\psi_{\alpha n}(c, 1)}{|\nu_{\alpha n}(c)|}\, \frac{\cos\{c\eta-\frac{1}{2}(n+\alpha+1)\pi\}}{(c\eta)^{\alpha+1}}\left\{1+O\left(\frac{1}{\eta}\right)\right\} \tag{A3.12}$$

for Re $\eta > 0$.

The prolate spheroidal wave functions $\{S_{m,m+n}(c, \eta)\}$ and the periodic Mathieu functions $\{Se_n(c, \eta)\}$ and $\{So_{n+1}(c, \eta)\}$ are all related simply to special cases of the spheroidal functions $\{\psi_{\alpha n}(c, \eta)\}$. From the integral equation for the prolate spheroidal wave functions as derived by Morse and Feshbach,† for example,

$$\frac{2(2m+n)!\, i^{m+n}}{c^m n!\, \lambda_{m,m+n}(c)} \frac{S_{m,m+n}(c,\eta)}{(1-\eta^2)^{m/2}} = \int_{-1}^{1} e^{ic\eta t}(1-t^2)^m \frac{S_{m,m+n}(c,t)}{(1-t^2)^{m/2}}\, dt, \tag{A3.13}$$

and from the integral equations for the even and odd periodic Mathieu functions as derived by Stratton,‡

$$\frac{(2\pi)^{\frac{1}{2}} i^n}{g_{e,n}(c^2)} Se_n(c,\eta) = \int_{-1}^{1} e^{ic\eta t}(1-t^2)^{-\frac{1}{2}} Se_n(c,t)\, dt, \tag{A3.14}$$

$$\frac{(2\pi)^{\frac{1}{2}} i^n}{c g_{o,n+1}(c^2)} \frac{So_{n+1}(c,\eta)}{(1-\eta^2)^{\frac{1}{2}}} = \int_{-1}^{1} e^{ic\eta t}(1-t^2)^{\frac{1}{2}} \frac{So_{n+1}(c,t)}{(1-t^2)^{\frac{1}{2}}}\, dt, \tag{A3.15}$$

with the characteristic numbers expressed in terms of Blanch's joining factors§ $\{g_{e,n}(c^2)\}$ and $\{g_{o,n+1}(c^2)\}$, it is evident by comparison with (A3.4) that

$$\psi_{\alpha n}(c,\eta) = \begin{cases} \dfrac{S_{m,m+n}(c,\eta)}{(1-\eta^2)^{m/2}}, & \alpha = m = 0, 1, 2, 3, \ldots \\ Se_n(c,\eta), & \alpha = -\frac{1}{2} \\ \dfrac{So_{n+1}(c,\eta)}{(1-\eta^2)^{\frac{1}{2}}}, & \alpha = +\frac{1}{2}, \end{cases} \tag{A3.16}$$

to within an arbitrary normalization factor.

The most complete treatment available on the theory of spheroidal functions is given by Meixner and Schäfke.‖

A3.1. Double orthogonality

An important property that will be established here is that the set of spheroidal functions $\{\psi_{\alpha n}(c,\eta)\}$ of arbitrary real order $\alpha > -1$ is orthogonal with weight factor $|1-\eta^2|^\alpha$ on two different intervals simultaneously. This double orthogonality property was first recognized by Slepian and Pollak¶

† P. M. Morse and H. Feshbach; *loc. cit.*, pp. 1505–06.

‡ J. A. Stratton. *Electromagnetic theory*, McGraw-Hill., New York, 1941; pp. 382 and 384.

§ National Bureau of Standards Computation Laboratory. *Tables relating to Mathieu functions*, Columbia University Press, N.Y. 1951; introduction by G. Blanch.

‖ J. Meixner and F. W. Schäfke. *Mathieusche Funktionen und Sphäroidfunktionen*, Springer-Verlag, Berlin, 1954.

¶ D. Slepian and H. O. Pollak, *loc. cit.*

for the case $\alpha = 0$; i.e. for the prolate spheroidal wave functions of order zero. When generalized to arbitrary values of α it will be seen in section A3.3 to lead to the existence of a third kind of characteristic numbers for the spheroidal functions.

Double orthogonality will be established from the Sturm–Liouville theory of second order differential equations. The solutions of all Sturm–Liouville systems, consisting of the differential equation and boundary conditions,

$$(pu')'+(\lambda w-q)u = 0, \qquad puu'\big|_{\eta=\eta_1} = puu'\big|_{\eta=\eta_2} = 0, \tag{A3.17}$$

are known from Hilbert theory† to form an infinite set of functions that is orthogonal and complete with respect to all square-integrable functions on the finite interval (η_1, η_2). After changing the spheroidal functions into functions that satisfy a Sturm–Liouville differential equation their double orthogonality property will be obtained.

The differential equation for the functions $\{S_{\alpha,\alpha+n}(c, \eta)\}$ defined by

$$\psi_{\alpha n}(c, \eta) = \frac{S_{\alpha,\alpha+n}(c, \eta)}{(1-\eta^2)^{\alpha/2}} \tag{A3.18}$$

is found from (A3.1) to be a singular Sturm–Liouville equation,

$$\{(1-\eta^2)S'_{\alpha,\alpha+n}(c, \eta)\}'+\left\{b_{\alpha n}+\alpha(\alpha+1)-c^2\eta^2-\frac{\alpha^2}{1-\eta^2}\right\}S_{\alpha,\alpha+n}(c, \eta) = 0. \tag{A3.19}$$

Multiplication by the complex conjugate $S^*_{\alpha,\alpha+p}(c, \eta)$ and subtraction of the product equation obtained when $S_{\alpha,\alpha+n}(c, \eta)$ and $S^*_{\alpha,\alpha+p}(c, \eta)$ are interchanged gives

$$\frac{d}{d\eta}\{(1-\eta^2)(S'_{\alpha,\alpha+n}S^*_{\alpha,\alpha+p}-S_{\alpha,\alpha+n}S^{*\prime}_{\alpha,\alpha+p})\} = (b_{\alpha p}-b_{\alpha n})S_{\alpha,\alpha+n}S^*_{\alpha,\alpha+p}. \tag{A3.20}$$

Integrating with respect to η over the real interval (η_1, η_2) and then changing to $\psi_{\alpha n}(c, \eta)$,

$$(1-\eta^2)\,|1-\eta^2|^\alpha\,\{\psi_{\alpha n}'(c, \eta)\psi_{\alpha p}(c, \eta)-\psi_{\alpha n}(c, \eta)\psi_{\alpha p}'(c, \eta)\}\Big|_{\eta=\eta_1}^{\eta_2}$$
$$= (b_{\alpha p}-b_{\alpha n})\int\limits_{\eta_1}^{\eta_2}\psi_{\alpha n}(c, \eta)\psi_{\alpha p}(c, \eta)\,|1-\eta^2|^\alpha\,d\eta. \tag{A3.21}$$

The characteristic numbers for all Sturm–Liouville systems (except those with periodic boundary conditions) are known to be distinct,‡ hence the spheroidal functions will be orthogonal on any interval (η_1, η_2) for which the left-hand side of (A3.21) is zero.

† H. Margenau and G. M. Murphy. *The mathematics of physics and chemistry*, Van Nostrand, New York, 1943; pp. 253–64.

‡ R. Courant and D. Hilbert, *loc. cit.*, pp. 293–4.

On the interval $(-1, 1)$ the left-hand side of (A3.21) is zero for all $\alpha > -1$ because the factor $(1-\eta^2)\,|1-\eta^2|^\alpha$ vanishes at both end points and because the spheroidal functions $\{\psi_{\alpha n}(c, \eta)\}$ and their derivatives are finite everywhere. Hence it follows that

$$\int_{-1}^{1} \psi_{\alpha n}(c, \eta)\psi_{\alpha p}(c, \eta)(1-\eta^2)^\alpha \, d\eta = 0, \qquad p \neq n; \qquad \text{(A3.22)}$$

i.e. the spheroidal functions are orthogonal on $(-1, 1)$ with weight factor $(1-\eta^2)^\alpha$.

From the asymptotic behaviour (A3.12) of $\psi_{\alpha n}(c, \eta)$ a simple calculation shows that for real η as $\eta \to \infty$,

$$(1-\eta^2)\,|1-\eta^2|^\alpha\{\psi_{\alpha n}'(c, \eta)\psi_{\alpha p}(c, \eta)-\psi_{\alpha n}(c, \eta)\psi_{\alpha p}'(c, \eta)\}$$
$$= \frac{\psi_{\alpha n}(c, 1)\psi_{\alpha p}(c, 1)4^{\alpha+1}\Gamma^2(\alpha+1)}{|\nu_{\alpha n}(c)|\,|\nu_{\alpha p}(c)|\,c^{2\alpha+1}}\left\{\sin \tfrac{1}{2}(p-n)\pi + O\left(\frac{1}{\eta}\right)\right\}. \qquad \text{(A3.23)}$$

When $p-n$ is even the first term on the right-hand side is zero for all η. Then the left-hand side of (A3.21) will be zero for $\eta_1 = 1$ and $\eta_2 = \infty$. Hence, on the interval $(1, \infty)$,

$$\int_{1}^{\infty} \psi_{\alpha n}(c, \eta)\psi_{\alpha p}(c, \eta)\,|1-\eta^2|^\alpha \, d\eta = 0, \qquad p-n \text{ even and } \neq 0. \qquad \text{(A3.24)}$$

Similarly on the interval $(-\infty, -1)$, because of symmetry of the functions. Combining the integrals on the three intervals $(-\infty, -1)$, $(-1, 1)$, and $(1, \infty)$,

$$\int_{-\infty}^{\infty} \psi_{\alpha n}(c, \eta)\psi_{\alpha p}(c, \eta)\,|1-\eta^2|^\alpha \, d\eta = 0, \qquad p \neq n; \qquad \text{(A3.25)}$$

i.e. the spheroidal functions are orthogonal on $(-\infty, \infty)$ with weight factor $|1-\eta^2|^\alpha$, including those for $p-n$ odd because the product $\psi_{\alpha n}(c, \eta)\psi_{\alpha p}(c, \eta)$ is then an odd function.

Thus it is concluded that the spheroidal functions $\{\psi_{\alpha n}(c, \eta)\}$ are orthogonal with weight factor $|1-\eta^2|^\alpha$ on the two intervals $(-1, 1)$ and $(-\infty, \infty)$ simultaneously.

A3.2. Normalization constants on $(-1, 1)$ and $(-\infty, \infty)$

Double orthogonality leads to two independent sets of normalization constants, one for each of the two intervals of orthogonality. One set can be chosen arbitrarily but the other must be determined from properties of the functions themselves. For consistency with the conventional theory of spheroidal functions the arbitrarily chosen set will be taken to be the one on the interval $(-1, 1)$. The orthonormality relation on $(-1, 1)$ can then be designated by

$$\int_{-1}^{1} \psi_{\alpha n}(c, \eta)\psi_{\alpha p}(c, \eta)(1-\eta^2)^\alpha \, d\eta = \Lambda_{\alpha n}(c)\, \delta_{np}, \tag{A3.26}$$

where $\{\Lambda_{\alpha n}(c)\}$ is a set of arbitrary positive numbers.

On the other interval $(-\infty, \infty)$ the normalization constants will be evaluated from two known properties of the spheroidal functions, namely, (i) their double orthogonality, and (ii) their completeness on $(-1, 1)$, as follows. Replacing $\psi_{\alpha p}(c, \eta)$ in the orthogonality integral (A3.25) by its integral representation (A3.4) and interchanging the order of integration,

$$\int_{-\infty}^{\infty} \psi_{\alpha n}(c, \eta)\psi_{\alpha p}(c, \eta)\, |1-\eta^2|^\alpha \, d\eta$$

$$= \int_{-1}^{1} \left\{ \frac{1}{\nu_{\alpha p}(c)} \int_{-\infty}^{\infty} e^{ic\eta t}\, |1-\eta^2|^\alpha \, \psi_{\alpha n}(c, \eta)\, d\eta \right\} (1-t^2)^\alpha \psi_{\alpha p}(c, t)\, dt. \tag{A3.27}$$

This must be zero for all $p \neq n$. Hence, from the other orthogonality integral (A3.22) it is evident that the bracketed expression must be orthogonal with weight factor $(1-t^2)^\alpha$ on $(-1, 1)$ to all of the functions $\{\psi_{\alpha p}(c, t)\}$ except the one for $p = n$. From the completeness property of the spheroidal functions on $(-1, 1)$, then, it is concluded that the bracketed expression for $p = n$ must be proportional to the nth function $\psi_{\alpha n}(c, t)$ everywhere on $(-1, 1)$:

$$\frac{1}{\nu_{\alpha n}(c)} \int_{-\infty}^{\infty} e^{ic\eta t}\, |1-\eta^2|^\alpha \, \psi_{\alpha n}(c, \eta)\, d\eta \equiv \gamma_{\alpha n}(c)\psi_{\alpha n}(c, t), \qquad -1 < t < 1, \tag{A3.28}$$

where $\gamma_{\alpha n}(c)$ denotes the proportionality constant. Thus, the orthonormality relation on $(-\infty, \infty)$ is

$$\int_{-\infty}^{\infty} \psi_{\alpha n}(c, \eta)\psi_{\alpha p}(c, \eta)\, |1-\eta^2|^\alpha \, d\eta = \gamma_{\alpha n}(c)\Lambda_{\alpha n}(c)\, \delta_{np}; \tag{A3.29}$$

i.e. the normalization constants on $(-\infty, \infty)$ are proportional to the arbitrarily chosen normalization constants $\{\Lambda_{\alpha n}(c)\}$ on $(-1, 1)$, where the proportionality constants are the real positive numbers $\{\gamma_{\alpha n}(c)\}$ defined by (A3.28).

A3.3. Characteristic numbers $\{\gamma_{\alpha n}(c)\}$

The new numbers $\{\gamma_{\alpha n}(c)\}$ that arise from the double orthogonality property are characteristic of the functions $\{\psi_{\alpha n}(c, \eta)\}$ in much the same sense as the characteristic numbers $\{b_{\alpha n}(c)\}$ and $\{\nu_{\alpha n}(c)\}$ of the spheroidal differential and integral equations (A3.1) and (A3.4), respectively, even though their defining equation (A3.28) is not an integral equation in any ordinary sense.

They can be obtained from the characteristic numbers of a true integral equation by using the complex conjugate of (A3.4) to represent $\psi_{\alpha n}(c, \eta)$ in (A3.28) and then interchanging the order of integration. The integration on η becomes an integral representation for the Neumann function, resulting in a new integral equation for the spheroidal functions,

$$|\nu_{\alpha n}(c)|^2\{\gamma_{\alpha n}(c)-1\}\psi_{\alpha n}(c,t) = -2^{\alpha+\frac{1}{2}}\Gamma(\alpha+1)\Gamma(\tfrac{1}{2})\int_{-1}^{1}\frac{Y_{-(\alpha+\frac{1}{2})}(c\,|t-x|)}{(c\,|t-x|)^{\alpha+\frac{1}{2}}}(1-x^2)^{\alpha}\psi_{\alpha n}(c,x)\,dx,$$
$$-1<t<1. \quad \text{(A3.30)}$$

Its range of validity as an integral equation is limited to $-1<\alpha<0$, because of the fact that its kernel contains nonintegrable singularities at $x=\pm 1$ for $\alpha \leq -1$ and at $x=t$ for $0<\alpha\neq m$. But even for higher orders $\alpha \geq 0$, where it is no longer valid as an integral equation, computational formulas for the characteristic numbers $\{\gamma_{\alpha n}(c)\}$ have been obtained from it that are found numerically to provide an accurate continuation from the range $-1<\alpha<0$.

For all nonnegative integral orders $\alpha = m = 0, 1, 2, 3, \ldots$ the singularity at $x = t$ is again integrable, leading to a closed-form expression for $\{\gamma_{mn}(c)\}$,

$$\gamma_{mn}(c) = 1-(-1)^m+(-1)^m\frac{2\pi}{|\nu_{mn}(c)|^2}\times$$
$$\times\left[\frac{\sum_{p=0}^{m}\frac{\binom{m}{p}}{c^{2p+1}}\frac{d^{2p}}{dt^{2p}}\{(1-t^2)^m\psi_{mn}(c,t)\}}{\psi_{mn}(c,t)}\right], \quad -1<t<1. \quad \text{(A3.31)}$$

Evaluating it at $t = 0$ one obtains the simple result

$$\gamma_{mn}(c) = 1-(-1)^m+(-1)^m\frac{2\pi}{|\nu_{mn}(c)|^2}\sum_{p=0}^{m}\frac{\binom{m}{p}}{c^{2p+1}}\beta_{2p}(c), \quad \text{(A3.32)}$$

where

$$\beta_{2p}(c) = \begin{cases} 0, & p<0 \\ 1, & p=0 \\ \{(2p-1)(2p-2-2m)-b_{mn}(c)\}\beta_{2p-2}(c)+ & \\ \qquad +(2p-2)(2p-3)c^2\beta_{2p-4}(c), & p>0. \end{cases}$$

Thus, for all nonnegative integral orders $\alpha = m$ the new characteristic numbers $\{\gamma_{\alpha n}(c)\}$ can be expressed directly in terms of the characteristic numbers $\{b_{mn}(c)\}$ and $\{\nu_{mn}(c)\}$ of the conventional differential and integral equations for $\{\psi_{mn}(c, \eta)\}$.

The method used here to obtain the defining equation (A3.28) for the new numbers $\{\gamma_{\alpha n}(c)\}$ was introduced originally by the author[3] to obtain a result similar to (A3.31) for the special cases of the prolate and oblate spheroidal wave functions of order $\alpha = m$. That result has since been confirmed in an elegant and quite different way by Meixner,† who generalized it to include all solutions of the differential equation (A3.1) of integral order $\alpha = m$ for arbitrary complex values of the index n.

Another special case occurs for certain half-integral orders. The new numbers $\{\gamma_{\alpha n}(c)\}$ for $\alpha = \pm\frac{1}{2}$ can be expressed directly in terms of Blanch's‡ joining factors $\{f_{e,n}(c^2)\}$ and $\{f_{o,n+1}(c^2)\}$ of the periodic Mathieu functions,

$$\gamma_{\alpha n}(c) = \begin{cases} 1+f_{e,n}(c^2), & \alpha = -\frac{1}{2} \\ 1+f_{o,n+1}(c^2), & \alpha = +\frac{1}{2}. \end{cases} \tag{A3.33}$$

Blanch's other joining factors $\{g_{e,n}(c^2)\}$ and $\{g_{o,n+1}(c^2)\}$ are seen from integral equations (A3.14) and (A3.15) to be inversely proportional to the characteristic numbers $\{|\nu_{\alpha n}(c)|\}$ for $\alpha = \pm\frac{1}{2}$. Thus, all of the joining factors of the periodic Mathieu functions can be obtained from the characteristic numbers of the spheroidal integral equations (A3.4) and (A3.30) for $\alpha = \pm\frac{1}{2}$.

For completely arbitrary values of $\alpha > -1$ the characteristic numbers $\{\gamma_{\alpha n}(c)\}$ can be computed in at least three different ways. One is from the defining equation (A3.28). Another is from the orthonormality relation (A3.29) obtained from it. And the third is from the integral equation (A3.30), also obtained from it. The simplest appears to be from the defining equation itself by evaluating it at some convenient value of t in the interval $(-1, 1)$. At $t = 0$ one obtains, for even $n = 2r$,

$$\gamma_{\alpha,2r}(c) = 1+\frac{2\int_1^\infty (\eta^2-1)^\alpha \psi_{\alpha,2r}(c, \eta)\, d\eta}{\nu_{\alpha,2r}(c)\psi_{\alpha,2r}(c, 0)}. \tag{A3.34a}$$

For odd $n = 2r+1$ the ratio becomes indeterminate at $t = 0$ because $\psi_{\alpha,2r+1}(c, \eta)$ is an odd function of η. It can be made determinate by applying L'Hospital's rule, giving

$$\gamma_{\alpha,2r+1}(c) = 1+\frac{i2c\int_1^\infty \eta(\eta^2-1)^\alpha \psi_{\alpha,2r+1}(c, \eta)\, d\eta}{\nu_{\alpha,2r+1}(c)\psi'_{\alpha,2r+1}(c, 0)}. \tag{A3.34b}$$

† J. Meixner. Einige eigenshaften der sphäroidfunktionen. *Arch. Math.*, **20**, pp. 274–8; 1969.

‡ NBS Computation Laboratory, *loc. cit.*

This representation will be used in section A3.7 to develop explicit computational formulas for $\{\gamma_{\alpha n}(c)\}$ that are valid for all $\alpha > -1$.

A3.4. Double completeness

The spheroidal functions $\{\psi_{\alpha n}(c, \eta)\}$ are related to the functions $\{S_{\alpha,\alpha+n}(c, \eta)\}$ that satisfy the Sturm-Liouville equation (A3.19) and the boundary condition that they must remain finite at their singular points $\eta = \pm 1$. Hence it follows from the Hilbert theory of Sturm–Liouville systems that they are complete on $(-1, 1)$ with respect to all functions that are square-integrable with weight factor $(1-\eta^2)^\alpha$. That completeness property will now be used to show that they are also complete on $(-\infty, \infty)$ with respect to the class of aperture-limited functions $F_\alpha(c\eta)$ of order α defined by

$$F_\alpha(c\eta) = \int_{-1}^{1} e^{ic\eta t}(1-t^2)^\alpha f(t)\, dt \tag{A3.35}$$

for real $\alpha > -1$. The only restriction on the function $f(t)$ is that it be square-integrable on $(-1, 1)$ with weight factor $(1-t^2)^\alpha$.

The function $f(t)$ can be expanded on $(-1, 1)$ in an infinite series of spheroidal functions that converges in the mean with weight factor $(1-t^2)^\alpha$,

$$f(t) = \sum_{n=0}^{\infty} \frac{a_n}{\nu_{\alpha n}(c)} \psi_{\alpha n}(c, t), \qquad -1 < t < 1, \tag{A3.36}$$

where

$$a_n = \frac{\nu_{\alpha n}(c)}{\Lambda_{\alpha n}(c)} \int_{-1}^{1} f(t)\psi_{\alpha n}(c, t)(1-t^2)^\alpha\, dt. \tag{A3.37}$$

Hence for all real values of η the absolute value of the difference

$$\left|F_\alpha(c\eta) - \sum_{n=0}^{\infty} a_n \psi_{\alpha n}(c, \eta)\right| = \left|\int_{-1}^{1} e^{ic\eta t}(1-t^2)^\alpha \left\{f(t) - \sum_{n=0}^{\infty} \frac{a_n}{\nu_{\alpha n}} \psi_{\alpha n}(c, t)\right\} dt\right|$$

$$\leqslant \int_{-1}^{1} (1-t^2)^\alpha \left|f(t) - \sum_{n=0}^{\infty} \frac{a_n}{\nu_{\alpha n}} \psi_{\alpha n}(c, t)\right| dt \tag{A3.38}$$

is exactly zero. It is concluded, therefore, that $F_\alpha(c\eta)$ can be represented by

$$F_\alpha(c\eta) = \sum_{n=0}^{\infty} a_n \psi_{\alpha n}(c, \eta), \qquad -\infty \leqslant \eta \leqslant \infty, \tag{A3.39}$$

where the series converges to $F_\alpha(c\eta)$ not just in the mean but uniformly on $(-\infty, \infty)$. Thus the functions $\{\psi_{\alpha n}(c, \eta)\}$ are complete on $(-\infty, \infty)$

with respect to all aperture-limited functions of order α, in addition to being complete on $(-1, 1)$ with respect to all square-integrable functions with weight factor $(1-\eta^2)^\alpha$.

A3.5. An extremal property

An important property of the prolate spheroidal wave functions of order zero was established by Landau and Pollak,† namely, that the first $N+1$ of them are the $N+1$ linearly independent aperture-limited functions of order zero that are most concentrated in the interval $(-1, 1)$. By the concentration of an aperture-limited function of order α will be meant the value of the ratio

$$R_\alpha(c) = \frac{\int_{-1}^{1} |F_\alpha(c\eta)|^2 (1-\eta^2)^\alpha \, d\eta}{\int_{-\infty}^{\infty} |F_\alpha(c\eta)|^2 |1-\eta^2|^\alpha \, d\eta}. \tag{A3.40}$$

It will now be shown, by quite a different and much more simple argument, that for *any* order $\alpha > -1$ the first $N+1$ spheroidal functions $\{\psi_{\alpha n}(c, \eta)\}$ are the $N+1$ linearly independent aperture-limited functions of order α that are most concentrated in the interval $(-1, 1)$.

The most general aperture-limited function $F_\alpha(c\eta)$ can be expanded in spheroidal functions as in (A3.39). Making use of the two orthonormality relations (A3.26) and (A3.29) the concentration ratio (A3.40) can be represented by

$$R_\alpha = \frac{\sum_{n=0}^{\infty} a_n a_n^* \Lambda_{\alpha n}}{\sum_{n=0}^{\infty} a_n a_n^* \gamma_{\alpha n} \Lambda_{\alpha n}}. \tag{A3.41}$$

Its value is determined entirely by the choice of expansion coefficients $\{a_n\}$ of the function $F_\alpha(c\eta)$. Hence a necessary condition for it to attain its largest value is that it be an extremum with respect to the real and the imaginary parts of each of the a_ns, or equivalently

$$\frac{\partial R_\alpha}{\partial a_p} = \frac{\partial R_\alpha}{\partial a_p^*} = 0, \qquad p = 0, 1, 2, 3, \ldots . \tag{A3.42}$$

Rewriting (A3.41) in the form

$$\sum_{n=0}^{\infty} a_n a_n^* (R_\alpha \gamma_{\alpha n} - 1) \Lambda_{\alpha n} = 0, \tag{A3.43}$$

† H. J. Landau and H. O. Pollak. Prolate spheroidal wave functions, Fourier analysis, and uncertainty—III. *Bell System Tech. J.*, **41**, pp. 1295–1336, July 1962.

then differentiating through with respect to a_p or $a_p{}^*$ and imposing condition (A3.42),

$$a_p{}^*(R_\alpha\gamma_{\alpha p}-1)\Lambda_{\alpha p} = a_p(R_\alpha\gamma_{\alpha p}-1)\Lambda_{\alpha p} = 0, \qquad p = 0, 1, 2, 3, \dots . \tag{A3.44}$$

The normalization constants $\{\Lambda_{\alpha p}(c)\}$ are never zero, hence either

$$a_p = 0, \tag{A3.45a}$$

or

$$R_\alpha\gamma_{\alpha p}-1 = 0. \tag{A3.45b}$$

The latter can be true for at most only one value of p, in which case the former must be true for all other values of p. It is clear, then, that an extremum of $R_\alpha(c)$ can occur when, and only when, $F_\alpha(c\eta)$ is proportional to one of the spheroidal functions $\{\psi_{\alpha n}(c, \eta)\}$. The value of $R_\alpha(c)$ at each of these extrema is, from (A3.45b),

$$R_{\alpha\text{ext}}(c) = \frac{1}{\gamma_{\alpha n}(c)}, \qquad n = 0, 1, 2, 3, \dots . \tag{A3.46}$$

The largest value possible is that for which $\gamma_{\alpha n}(c)$ is least, which is found to be $\gamma_{\alpha 0}(c)$. Hence the spheroidal function for $n = 0$ is the single most concentrated aperture-limited function of order α.

The spheroidal functions form a linearly independent set on $(-\infty, \infty)$, since they are orthogonal on that interval. Thus, the second most concentrated aperture-limited function of order α that is linearly independent of $\psi_{\alpha 0}(c, \eta)$ can be expanded in the set of functions $\{\psi_{\alpha n}(c, \eta)\}$ that remains after the most concentrated function $\psi_{\alpha 0}(c, \eta)$ has been removed; i.e. after setting a_0 equal to zero in (A3.39). Repeating the maximization process above leads once again to conditions (A3.45) but restricted now to $p=1, 2, 3, \dots$. The largest possible value is again that for which $\gamma_{\alpha p}(c)$ is least, which now is $\gamma_{\alpha 1}(c)$. Hence the spheroidal functions for $n = 0$ and 1 form the most concentrated *pair* of linearly independent aperture-limited functions of order α. By this process of systematically depleting the set of functions $\{\psi_{\alpha n}(c, \eta)\}$ it is concluded that the first $N+1$ of them are, indeed, the $N+1$ linearly independent aperture-limited functions of order α that are most concentrated in the interval $(-1, 1)$.

A3.6. A special case

For the special case in which $c = \frac{1}{2}q\pi$, $\alpha = 1$, $n = q-1$, with $q = 1, 2, 3, \dots$, an exact closed-form representation can be obtained for the spheroidal function and for each of its associated numbers. The spheroidal function itself can be obtained from the differential equation (A3.1) and the boundary condition of finiteness at $\eta = \pm 1$ by letting

$$\psi_{\alpha n}(c, \eta) = \frac{y_{\alpha n}(c, \eta)}{1-\eta^2}, \tag{A3.47}$$

which leads to the following differential equation for $y_{\alpha n}(c,\eta)$,

$$(1-\eta^2)y_{\alpha n}''-2(\alpha-1)\eta y_{\alpha n}'+\left(b_{\alpha n}+2-c^2\eta^2-\frac{4(\alpha-1)\eta^2}{1-\eta^2}\right)y_{\alpha n}=0. \quad \text{(A3.48)}$$

For $\alpha=1$ this reduces to

$$(1-\eta^2)y_{1n}''+(b_{1n}+2-c^2\eta^2)y_{1n}=0. \quad \text{(A3.49)}$$

If $c=\frac{1}{2}q\pi$ and $n=q-1$, and if the value of $b_{\alpha n}$ is chosen to be

$$b_{1,q-1}(\tfrac{1}{2}q\pi)=(\tfrac{1}{2}q\pi)^2-2, \quad \text{(A3.50)}$$

the differential equation (A3.49) becomes

$$y''_{1,q-1}+(\tfrac{1}{2}q\pi)^2y_{1,q-1}=0, \quad \text{(A3.51)}$$

whose solution is

$$y_{1,q-1}(\tfrac{1}{2}q\pi,\eta)=k_q\begin{Bmatrix}\cos\\ \sin\end{Bmatrix}\tfrac{1}{2}q\pi\eta, \quad \text{(A3.52)}$$

where k_q is an arbitrary scale factor. Hence it is concluded that

$$\psi_{1,q-1}(\tfrac{1}{2}q\pi,\eta)=k_q\frac{\left.\begin{matrix}\cos\\ \sin\end{matrix}\right\}\tfrac{1}{2}q\pi\eta}{1-\eta^2}, \qquad q\ \begin{matrix}\text{odd}\\ \text{even}\end{matrix}, \quad \text{(A3.53)}$$

from the fact that it satisfies the original differential equation (A3.1), that it is finite at $\eta=\pm 1$, and that it has exactly $q-1$ zeros in the interval $(-1,1)$.

The spheroidal function (A3.53) and the characteristic number (A3.50) of its differential equation were obtained earlier in a similar way by Flammer.† The same closed-form representation for the spheroidal function can be obtained also from the fact that it satisfies the integral equation (A3.4), as can be seen by direct evaluation of the following integral,

$$\int_{-1}^{1}e^{i(q\pi/2)\eta t}(1-t^2)\frac{\left.\begin{matrix}\cos\\ \sin\end{matrix}\right\}\tfrac{1}{2}q\pi t}{1-t^2}\,dt=i^{q-1}\frac{4}{q\pi}\,\frac{\left.\begin{matrix}\cos\\ \sin\end{matrix}\right\}\tfrac{1}{2}q\pi\eta}{1-\eta^2}. \quad \text{(A3.54)}$$

This is precisely the integral equation (A3.4) in the particular case of $\alpha=1$ and $c=\frac{1}{2}q\pi$, from which it is evident that its characteristic number must be

$$\nu_{1,q-1}(\tfrac{1}{2}q\pi)=i^{q-1}(4/q\pi). \quad \text{(A3.55)}$$

The normalization constant on $(-1,1)$ can be obtained in terms of the scale factor k_q by direct evaluation of the integral (A3.26),

$$\Lambda_{1,q-1}(\tfrac{1}{2}q\pi)=\tfrac{1}{2}k_q^2\,\mathrm{Cin}\,2q\pi, \quad \text{(A3.56)}$$

expressed in terms of the cosine integral defined by (A2.11). And by direct evaluation of the other normalization integral (A3.29) one finds that

$$\gamma_{1,q-1}(\tfrac{1}{2}q\pi)=2. \quad \text{(A3.57)}$$

† C. Flammer. *Spheroidal wave functions*, Stanford University Press, Stanford, California, 1957; pp. 24–25.

These results lead to an interesting indication of the way in which the value of the integral in the definition (A3.28) of the new characteristic numbers $\{\gamma_{\alpha n}(c)\}$ can behave outside the open interval $-1 < t < 1$ within which the identity must hold. For q odd, for example, the integral can be evaluated in closed form,

$$i^{q-1}\frac{q\pi}{4}\int_{-\infty}^{\infty} e^{i(q\pi/2)\eta t}\,|1-\eta^2|\frac{\cos\frac{1}{2}q\pi\eta}{1-\eta^2}\,d\eta$$

$$= 2\frac{\cos\frac{1}{2}q\pi t}{1-t^2} - i^{q-1}\tfrac{1}{2}\pi\{\delta(t-1)+\delta(t+1)\}, \quad \text{(A3.58)}$$

which agrees exactly with the right-hand side of (A3.28) within the open interval $-1 < t < 1$. It does not agree at the endpoints $t = \pm 1$, however, as evidenced by the delta function singularities that appear there.

A3.7. Computational formulas

Explicit formulas for use in computing the spheroidal functions and their associated numbers will be developed here.

(i) *The spheroidal functions* $\{\psi_{\alpha n}(c, \eta)\}$ *and characteristic numbers* $\{b_{\alpha n}(c)\}$

On the interval $(-1, 1)$ the Gegenbauer polynomials $\{T_k^{\alpha}(\eta)\}$ form a complete set with respect to all functions that are square-integrable with weight factor $(1-\eta^2)^{\alpha}$, and they are orthogonal with weight factor $(1-\eta^2)^{\alpha}$:

$$\int_{-1}^{1}(1-\eta^2)^{\alpha}T_k^{\alpha}(\eta)T_l^{\alpha}(\eta)\,d\eta = \frac{2\Gamma(k+2\alpha+1)}{(2k+2\alpha+1)k!}\,\delta_{kl}. \quad \text{(A3.59)}$$

They are even or odd functions according as k is even or odd. Hence the spheroidal functions can be expanded in them,

$$\psi_{\alpha n}(c, \eta) = \sum_{k=0,1}^{\infty}{}' d_k(c \mid \alpha n)T_k^{\alpha}(\eta), \quad \text{(A3.60)}$$

where the prime denotes summation over only even or odd values of k according as n is even or odd, respectively. The spheroidal function $\psi_{\alpha n}(c, \eta)$ becomes proportional to $T_n^{\alpha}(\eta)$ as $c \to 0$, from which it follows that

$$d_k(c \mid \alpha n) \to 0 \quad \text{for} \quad k \neq n. \quad \text{(A3.61)}$$

The expansion (A3.60) converges rapidly everywhere inside of the interval $(-1, 1)$ but outside it converges more and more slowly with increasing values of η. Furthermore it requires special consideration for the case $\alpha = -\frac{1}{2}$ because $T_k^{-\frac{1}{2}}(\eta)$ is infinite for $k = 0$. An alternative expansion is available that has the advantage of representing $\psi_{\alpha n}(c, \eta)$ uniformly on the entire real axis $(-\infty, \infty)$ for all $\alpha > -1$. It can be obtained from the integral representation (A3.4) by substituting the expansion (A3.60) for $\psi_{\alpha n}(c, \eta)$ inside the integral, integrating term by term, and then making use of the transform (5.10):

$$\psi_{\alpha n}(c, \eta) = \frac{i^n(2\pi)^{\frac{1}{2}}}{\nu_{\alpha n}(c)} \sum_{k=0,1}^{\infty}{}' a_k(c \mid \alpha n) \frac{J_{k+\alpha+\frac{1}{2}}(c\eta)}{(c\eta)^{\alpha+\frac{1}{2}}}, \tag{A3.62}$$

where the coefficients $\{a_k(c \mid \alpha n)\}$ are related to $\{d_k(c \mid \alpha n)\}$ by

$$a_k(c \mid \alpha n) = i^{k-n} \frac{\Gamma(k+2\alpha+1)}{k!} d_k(c \mid \alpha n). \tag{A3.63}$$

Hence *all of the computational formulas derived here will be expressed in terms of the expansion coefficients* $\{a_k(c \mid \alpha n)\}$. The coefficients $\{a_k(c \mid \alpha n)\}$ are real because $\{d_k(c \mid \alpha n)\}$ are real (the factor i^{k-n} is always real since $k-n$ is always an even integer). And when $c \to 0$,

$$a_k(c \mid \alpha n) \to 0 \quad \text{for} \quad k \neq n, \tag{A3.64}$$

by (A3.61).

For the special cases in which the value of α is a nonnegative integer $\alpha = m = 0, 1, 2, 3, \ldots$ the coefficients $\{a_k(c \mid \alpha n)\}$ in (A3.62) will agree with those tabulated by Stratton *et al.*† for the prolate spheroidal wave functions of order m provided that the functions are normalized such that

$$\sum_{k=0,1}^{\infty}{}' i^{k-n} a_k(c \mid \alpha n) = 1. \tag{A3.65}$$

And if the right-hand side of this normalization equation (A3.65) is replaced by $(\pi/2)^{\frac{1}{2}}$ the coefficients $\{a_k(c \mid \alpha n)\}$ for the special cases of $\alpha = -\frac{1}{2}$ and $+\frac{1}{2}$ will be related to Blanch's‡ Mathieu function coefficients $\{\mathrm{De}_k^{(n)}(c^2)\}$ and $\{\mathrm{Do}_{k+1}^{(n+1)}(c^2)\}$, respectively, by

$$a_k(c \mid -\tfrac{1}{2}, n) = i^{k-n}(\pi/2)^{\frac{1}{2}} \mathrm{De}_k^{(n)}(c^2), \tag{A3.66}$$

$$a_k(c \mid +\tfrac{1}{2}, n) = i^{k-n}(\pi/2)^{\frac{1}{2}}(k+1)\mathrm{Do}_{k+1}^{(n+1)}(c^2). \tag{A3.67}$$

A three-term recursion relation among the coefficients $\{a_k(c \mid \alpha n)\}$ can be obtained by substituting the Bessel function expansion (A3.62) for $\psi_{\alpha n}(c, \eta)$ in the differential equation (A3.1) and then using the recursion relations among Bessel functions,

$$A_{k+2}a_{k+2} + (b_{\alpha n} - B_k)a_k + C_{k-2}a_{k-2} = 0, \tag{A3.68}$$

† J. A. Stratton, P. M. Morse, L. J. Chu, J. D. C. Little, and F. J. Corbató. *Spheroidal wave functions*, The MIT Press, Cambridge, Massachusetts, 1956.

‡ NBS Computation Laboratory, *loc. cit.*

where
$$A_k = \frac{k(k-1)}{(2k+2\alpha-1)(2k+2\alpha+1)}c^2,$$
$$B_k = k(k+2\alpha+1)+\frac{2k^2+2k(2\alpha+1)+2\alpha-1}{(2k+2\alpha-1)(2k+2\alpha+3)}c^2,$$
$$C_k = \frac{(k+2\alpha+1)(k+2\alpha+2)}{(2k+2\alpha+1)(2k+2\alpha+3)}c^2.$$

It can be rewritten as a continued fraction for the ratio of two succeeding values of a_k in two different ways, one in terms of increasing values of the subscript,
$$\frac{a_k}{a_{k-2}} = -\frac{C_{k-2}}{b_{\alpha n}-B_k+A_{k+2}(a_{k+2}/a_k)}, \tag{A3.69}$$
and the other in terms of decreasing values,
$$\frac{a_{k-2}}{a_k} = -\frac{A_k}{b_{\alpha n}-B_{k-2}+C_{k-4}(a_{k-4}/a_{k-2})}. \tag{A3.70}$$
Each must be the reciprocal of the other, which is a condition that will be satisfied for only a certain infinite set of discrete values of the parameter $b_{\alpha n}$. These are the characteristic numbers of the differential equation. The set of coefficients $\{a_k(c \mid \alpha n)\}$ associated with each of these characteristic numbers determines the corresponding spheroidal function $\psi_{\alpha n}(c, \eta)$, to within a constant factor, by (A3.62).

The best method available for computing these characteristic numbers $\{b_{\alpha n}(c)\}$, and simultaneously the expansion coefficients $\{a_k(c \mid \alpha n)\}$, is an iterative procedure that was used by Bouwkamp† for the prolate spheroidal wave functions of order zero and by Blanch‡ for the Mathieu functions. But first there are two additional restrictions that must be imposed on the recursion relation (A3.68). For $k < 0$ the expansion coefficients must be zero, according to a result of L. Gegenbauer's§ in which he showed that any analytic function can be represented within its radius of analyticity by a series of the form (A3.62). And an indeterminacy of A_k, B_k, and C_k exists at $\alpha = \pm\frac{1}{2}$ for $k = 0$ and 1 that must be resolved. Therefore the recursion relation (A3.68) must be supplemented by two more conditions,
$$a_k = 0 \quad \text{for} \quad \mathrm{k} < 0, \tag{A3.71}$$

† C. J. Bouwkamp. Theoretische en numerieke behandeling van de buiging door een ronde opening. Dissertation, University of Groningen, 1941 (English trans: *IEEE Trans. Antennas Propagat.*, **AP-18,** pp. 152–76, March 1970). Also; On spheroidal wave functions of order zero. *J. Math. and Phys.* **26,** pp. 79–93, January 1947.

‡ G. Blanch. On the computation of Mathieu functions. *J. Math. and Phys.* **25,** pp. 1–20, January 1946.

§ G. N. Watson, *Theory of Bessel functions,* Cambridge University Press, 1962; section 16.13.

and

$$A_0 = A_1 = 0, \quad B_0 = \frac{c^2}{2\alpha+3}, \quad B_1 = 2\alpha+2+\frac{3c^2}{2\alpha+5}, \quad C_0 = \frac{2\alpha+2}{2\alpha+3}c^2, \tag{A3.72}$$

the latter of which resolves the indeterminacy.

Once the characteristic numbers $\{b_{\alpha n}(c)\}$ have been determined the ratio a_k/a_{k-2} can also be determined for all k, either from (A3.69) or (A3.70). The coefficients $\{a_k(c \mid \alpha n)\}$ can then be computed from these ratios in terms of the value chosen for the leading coefficient a_0 or a_1. The particular value chosen for a_0 or a_1 determines the scale factors by which the spheroidal functions are normalized, and conversely. But no agreement exists among authors on the choice of these scale factors. It is for this reason that *all formulas pertaining to spheroidal functions in this treatise have been expressed in terms of arbitrary normalization constants on the interval* $(-1, 1)$. For the numerical results shown in the next section the normalization constants on $(-1, 1)$ will be chosen such that the normalization constants on $(-\infty, \infty)$ all have the value unity.

The value of the spheroidal functions, and of their first derivatives, at the origin $\eta = 0$ appear in the computational formulas for the other two kinds of characteristic numbers. They are found from (A3.60) to be

$$\psi_{\alpha,2r}(c, 0) = \frac{(-1)^r}{2^\alpha \pi^{\frac{1}{2}}} \sum_{l=0}^{\infty} \frac{\Gamma(l+\frac{1}{2})}{\Gamma(l+\alpha+1)} a_{2l}(c \mid \alpha, 2r), \tag{A3.73}$$

$$\psi'_{\alpha,2r+1}(c, 0) = \frac{(-1)^r}{2^{\alpha-1}\pi^{\frac{1}{2}}} \sum_{l=0}^{\infty} \frac{\Gamma(l+\frac{3}{2})}{\Gamma(l+\alpha+1)} a_{2l+1}(c \mid \alpha, 2r+1). \tag{A3.74}$$

(ii) *The characteristic numbers* $\{\nu_{\alpha n}(c)\}$

The characteristic numbers $\{\nu_{\alpha n}(c)\}$ of the integral equation (A3.4) can be computed by evaluating the Bessel function expansion (A3.62), or its derivative, at $\eta = 0$. As η approaches zero for even $n = 2r$ the Bessel functions in (A3.62) cause all terms to vanish except the one for $k = 0$, hence

$$\nu_{\alpha,2r}(c) = \frac{(-1)^r \pi^{\frac{1}{2}} a_0(c \mid \alpha, 2r)}{2^\alpha \Gamma(\alpha+\frac{3}{2}) \psi_{\alpha,2r}(c, 0)}. \tag{A3.75a}$$

For odd $n = 2r+1$ the spheroidal function vanishes at $\eta = 0$, so (A3.62) must first be differentiated with respect to η before letting η approach zero. All terms then vanish except the one for $k = 1$, hence

$$\nu_{\alpha,2r+1}(c) = ic\frac{(-1)^r \pi^{\frac{1}{2}} a_1(c \mid \alpha, 2r+1)}{2^{\alpha+1}\Gamma(\alpha+\frac{5}{2}) \psi'_{\alpha,2r+1}(c, 0)}. \tag{A3.75b}$$

(iii) *The characteristic numbers* $\{\gamma_{\alpha n}(c)\}$

Computational formulas for the third kind of characteristic numbers $\{\gamma_{\alpha n}(c)\}$ will be developed here from their representation (A3.34). The integrals in (A3.34) can be evaluated in terms of the expansion coefficients $\{a_k(c \mid \alpha n)\}$ and certain integrals of Bessel functions by substituting the expansion (A3.62) for $\psi_{\alpha n}(c, \eta)$ and integrating term by term,

$$\gamma_{\alpha,2r}(c) = 1+\frac{2(2\pi)^{\frac{1}{2}}}{|\nu^2_{\alpha,2r}(c)\psi_{\alpha,2r}(c,0)|}\sum_{p=0}^{\infty} a_{2p}(c \mid \alpha, 2r)I_{\alpha,2p}(c), \tag{A3.76a}$$

$$\gamma_{\alpha,2r+1}(c) = 1+\frac{2(2\pi)^{\frac{1}{2}}}{|\nu^2_{\alpha,2r+1}\psi'_{\alpha,2r+1}(c,0)|}\sum_{p=0}^{\infty} a_{2p+1}(c \mid \alpha, 2r+1)K_{\alpha,2p}(c), \tag{A3.76b}$$

where $I_{\alpha k}(c)$ and $K_{\alpha k}(c)$ denotes the integrals

$$I_{\alpha k}(c) = \int_1^\infty (\eta^2-1)^\alpha \frac{J_{k+\alpha+\frac{1}{2}}(c\eta)}{(c\eta)^{\alpha+\frac{1}{2}}}\,d\eta, \tag{A3.77}$$

$$K_{\alpha k}(c) = \int_1^\infty (\eta^2-1)^\alpha \frac{J_{k+2+\alpha-\frac{1}{2}}(c\eta)}{(c\eta)^{\alpha-\frac{1}{2}}}\,d\eta. \tag{A3.78}$$

For $k = 0$ the integral $I_{\alpha k}(c)$ can be evaluated within the range $-1 < \alpha \leqslant 0$ by replacing the Bessel function in (A3.77) by its representation as an integral on $(-1, 1)$ and then interchanging the order of integration,

$$I_{\alpha 0}(c) = \frac{2^{-\alpha+\frac{1}{2}}}{\Gamma(\alpha+1)\Gamma(\frac{1}{2})}\int_0^1 (1-t^2)^\alpha \left\{\int_1^\infty (\eta^2-1)^\alpha \cos c\eta t\,d\eta\right\} dt. \tag{A3.79}$$

The inner integral is an integral representation for the Neumann function, hence

$$I_{\alpha 0}(c) = -\int_0^1 (1-t^2)^\alpha \frac{Y_{-(\alpha+\frac{1}{2})}(ct)}{(ct)^{\alpha+\frac{1}{2}}}\,dt$$

$$= -\frac{1}{\cos\alpha\pi}\left\{(\sin\alpha\pi)\int_0^1 (1-t^2)^\alpha \frac{J_{-(\alpha+\frac{1}{2})}(ct)}{(ct)^{\alpha+\frac{1}{2}}}\,dt + \int_0^1 (1-t^2)^\alpha \frac{J_{\alpha+\frac{1}{2}}(ct)}{(ct)^{\alpha+\frac{1}{2}}}\,dt\right\}, \tag{A3.80}$$

the latter being obtained from the definition of the Neumann function in terms of Bessel functions. Expanding the Bessel functions in power series about the origin and then integrating term by term gives

$$I_{\alpha 0}(c) = \frac{\Gamma(\alpha+1)}{2^{\alpha+\frac{3}{2}}\cos\alpha\pi}\sum_{l=0}^{\infty}\frac{(-1)^{l+1}}{l!}\times$$
$$\times\left\{\frac{(\sin\alpha\pi)\Gamma(l-\alpha)}{l!\,\Gamma(l-\alpha+\frac{1}{2})}\left(\frac{c}{2}\right)^{-2\alpha-1}+\frac{\Gamma(l+\frac{1}{2})}{\Gamma^2(l+\alpha+\frac{3}{2})}\right\}\left(\frac{c}{2}\right)^{2l}. \quad \text{(A3.81)}$$

For all even values of k the integrals $I_{\alpha k}(c)$ and $K_{\alpha k}(c)$ can now be evaluated within the range $-1 < \alpha \leqslant 0$ by means of recursion from this expression for $I_{\alpha 0}(c)$. From the recursion relations among Bessel functions the following recursion relationships among $I_{\alpha k}(c)$ and $K_{\alpha k}(c)$ can be established, valid for all $\alpha > -1$:

$$K_{\alpha,k-2}(c)+K_{\alpha k}(c) = (2k+2\alpha+1)I_{\alpha k}(c), \quad \text{(A3.82)}$$
$$K_{\alpha,k-2}(c)-K_{\alpha k}(c) = (2\alpha+1)I_{\alpha k}(c)+2cI'_{\alpha k}(c), \quad \text{(A3.83)}$$
$$K_{\alpha k}(c) = kI_{\alpha k}(c)-cI'_{\alpha k}(c), \quad \text{(A3.84)}$$
$$(k+2\alpha+3)I_{\alpha,k+2}(c)+cI'_{\alpha,k+2}(c) = kI_{\alpha k}(c)-cI'_{\alpha k}(c). \quad \text{(A3.85)}$$

The last relation (A3.85) is a differential equation for $I_{\alpha,k+2}(c)$ in terms of $I_{\alpha k}(c)$, the solution of which is

$$I_{\alpha,k+2}(c) = \frac{2k+2\alpha+3}{c^{k+2\alpha+3}}\int_0^c t^{k+2\alpha+2}I_{\alpha k}(t)\,dt-I_{\alpha k}(c). \quad \text{(A3.86)}$$

By repeated application of this recursion relation, starting from the power series representation (A3.81) for $I_{\alpha 0}(c)$, one can surmise the power series representation for $I_{\alpha,2p}(c)$, the validity of which is easily confirmed by mathematical induction. The qth derivative, obtained by differentiating term by term, is found to be

$$I^{(q)}_{\alpha,2p}(c) = \frac{\Gamma(\alpha+1)}{2^{q+\alpha+\frac{3}{2}}\cos\alpha\pi}\sum_{l=0}^{\infty}\frac{(-1)^{l+1}}{l!}\times$$
$$\times\left\{\frac{(\sin\alpha\pi)\Gamma(l-\alpha)(-l+\alpha+\frac{1}{2})_p(2l-2\alpha-q)_q}{(l+p)!\,\Gamma(l-\alpha+\frac{1}{2})}\left(\frac{c}{2}\right)^{-2\alpha-1}+\right.$$
$$\left.+\frac{\Gamma(l+\frac{1}{2})(-l)_p(2l+1-q)_q}{\Gamma(l+\alpha+\frac{3}{2})\Gamma(l+\alpha+\frac{3}{2}+p)}\right\}\left(\frac{c}{2}\right)^{2l-q}, \quad \text{(A3.87)}$$

where $(z)_p = z(z+1)(z+2)\ldots(z+p-1)$, $(z)_0 = 1$ denotes Pochhammer's symbol. Thus, the integrals $I_{\alpha,2p}(c)$ and $K_{\alpha,2p}(c)$ can be computed in the range $-1 < \alpha \leqslant 0$ from (A3.87) and (A3.84), respectively.

To compute $I_{\alpha,2p}(c)$, and thence $K_{\alpha,2p}(c)$, for orders beyond the range of integrability $-1 < \alpha \leqslant 0$ of (A3.80) the following recursion relation can be established, valid for all $\alpha > 0$,

$$I_{\alpha k}(c) = \left(\frac{k}{c^2} - \frac{1}{2k+2\alpha+1}\right) I_{\alpha-1,k}(c) - \frac{1}{2k+2\alpha+1} I_{\alpha-1,k+2}(c) - \frac{1}{c} I'_{\alpha-1,k}(c), \quad \text{(A3.88)}$$

again from the recursion relations among Bessel functions. It can be shown by direct substitution that the power series (A3.87) satisfies this recursion relation. Hence it follows by induction that the power series is valid for *all* $\alpha > -1$, even though the integral (A3.80) from which it was derived contains a nonintegrable singularity at $t = 0$ for all nonintegral values of α greater than zero.

As an indication of the rapidity of convergence of (A3.87) and (A3.76), and hence of the practicality of the formulas developed here, the series (A3.87) was found to converge to 16 significant figures in less than 23 terms for all of the computations shown in the next section, and the series (A3.76) to converge to 16 significant figures in less than 19 terms.

The series (A3.87) becomes indeterminate when α is an integer or a half-integer, but the indeterminacy can always be resolved by taking the limit. For the integer $\alpha = 0$, for example, the value obtained for $I_{\alpha 0}(c)$ is

$$I_{00}(c) = \frac{1}{c}\left(\frac{\pi}{2}\right)^{\frac{1}{2}} - \left(\frac{2}{\pi}\right)^{\frac{1}{2}} \sum_{l=0}^{\infty} (-1)^l \frac{c^{2l}}{(2l+1)(2l+1)!}, \quad \text{(A3.89)}$$

in agreement with the power series expansion of $(2/\pi)^{\frac{1}{2}}\{(\pi/2) - \operatorname{Si} c\}/c$ obtained by direct evaluation of the integral (A3.77). And for the half-integer $\alpha = -\frac{1}{2}$, for example, the value obtained for $I_{\alpha 0}(c)$ is

$$I_{-\frac{1}{2},0}(c) = \frac{1}{2\sqrt{\pi}} \sum_{l=0}^{\infty} (-1)^l \frac{\Gamma(l+\frac{1}{2})}{(l!)^3} \{3\psi(l+1) - \psi(l+\tfrac{1}{2}) - 2 \ln \tfrac{1}{2}c\}(\tfrac{1}{2}c)^{2l} \quad \text{(A3.90)}$$

in terms of the Digamma function $\psi(z)$, in agreement with the power series expansion of $-(\frac{1}{2}\pi)J_0(\frac{1}{2}c)Y_0(\frac{1}{2}c)$ obtained again by direct evaluation of the integral (A3.77). But in the special cases of integral orders $\alpha = m$ and of the half-integral orders $\alpha = \pm\frac{1}{2}$ there are more direct means available for evaluating $\{\gamma_{\alpha n}(c)\}$ than by use of the power series expansion for $I^{(q)}_{\alpha,2p}(c)$. For all integral orders the third kind of characteristic numbers $\{\gamma_{\alpha n}(c)\}$ has been expressed quite simply in terms of the other two kinds $\{b_{\alpha n}(c)\}$ and $\{\nu_{\alpha n}(c)\}$ by (A3.32). And for the half-integers $\alpha = \pm\frac{1}{2}$ the integrals $I_{\alpha,2p}(c)$ and $K_{\alpha,2p}(c)$ in (A3.76) can be evaluated in closed form in terms of Bessel functions. For $\alpha = -\frac{1}{2}$ the integral $I_{\alpha,2p}(c)$ is known to be†

$$I_{-\frac{1}{2},2p}(c) = -\tfrac{1}{2}\pi J_p(\tfrac{1}{2}c) Y_p(\tfrac{1}{2}c), \quad \text{(A3.91)}$$

† I. S. Gradshteyn and I. M. Ryzhik, *loc. cit.*, eqn. (6.552.6).

from which the corresponding integral $K_{\alpha,2p}(c)$ is found from recursion relation (A3.84) to be

$$K_{-\frac{1}{2},2p}(c) = -\tfrac{1}{4}\pi c\{J_p(\tfrac{1}{2}c)Y_{p+1}(\tfrac{1}{2}c)+J_{p+1}(\tfrac{1}{2}c)Y_p(\tfrac{1}{2}c)\}, \tag{A3.92}$$

and from which $I_{\alpha,2p}(c)$ and $K_{\alpha,2p}(c)$ for $\alpha = +\frac{1}{2}$ are found from recursion relations (A3.88) and (A3.84), respectively, to be

$$I_{\frac{1}{2},2p}(c) = \frac{\pi}{4(2p+1)}\bigg[J_p(\tfrac{1}{2}c)Y_p(\tfrac{1}{2}c)+J_{p+1}(\tfrac{1}{2}c)Y_{p+1}(\tfrac{1}{2}c)- -\frac{2p+1}{c}\{J_p(\tfrac{1}{2}c)Y_{p+1}(\tfrac{1}{2}c)+J_{p+1}(\tfrac{1}{2}c)Y_p(\tfrac{1}{2}c)\}\bigg], \tag{A3.93}$$

$$K_{\frac{1}{2},2p}(c) = \frac{\pi}{4}\bigg[J_p(\tfrac{1}{2}c)Y_p(\tfrac{1}{2}c)+J_{p+1}(\tfrac{1}{2}c)Y_{p+1}(\tfrac{1}{2}c)- -\frac{2p+2}{c}\{J_p(\tfrac{1}{2}c)Y_{p+1}(\tfrac{1}{2}c)+J_{p+1}(\tfrac{1}{2}c)Y_p(\tfrac{1}{2}c)\}\bigg]. \tag{A3.94}$$

The resulting series expansion (A3.76) of $\{\gamma_{\alpha n}(c)-1\}$ for $\alpha = -\frac{1}{2}$ and $+\frac{1}{2}$ is identical to Blanch's expansion of the joining factors $\{f_{e,n}(c^2)\}$ and $\{f_{o,n+1}(c^2)\}$, respectively, in (A3.33).

In summary, the characteristic numbers $\{\gamma_{\alpha n}(c)\}$ for all values of α greater than -1 can be computed from the expansions (A3.76), in which the value of the integrals $I_{\alpha,2p}(c)$ can be computed from the power series (A3.87) and in which the value of the integrals $K_{\alpha,2p}(c)$ can be computed also from (A3.87) by use of the relation (A3.84). For the special cases of $\alpha = \pm\frac{1}{2}$ (the Mathieu functions) the values of the integrals $I_{\alpha,2p}(c)$ and $K_{\alpha,2p}(c)$ are given simply by (A3.91) to (A3.94). And for the special cases of $\alpha = m = 0, 1, 2, 3, \ldots$ (the prolate spheroidal wave functions) the numbers $\{\gamma_{\alpha n}(c)\}$ can be evaluated more directly from (A3.32).

(iv) *The normalization constants* $\{\Lambda_{\alpha n}(c)\}$

The numbers $\{\Lambda_{\alpha n}(c)\}$ that denote the normalization constants on the interval $(-1, 1)$ will be evaluated directly from their definition (A3.26) for $p = n$. Expanding the spheroidal function in Gegenbauer polynomials as in (A3.60) and then making use of their orthogonality relation (A3.59) one obtains

$$\Lambda_{\alpha n}(c) = \sum_{k=0,1}^{\infty}{}' \frac{k!}{(k+\alpha+\frac{1}{2})\Gamma(k+2\alpha+1)} a_k^2(c \mid \alpha n) \tag{A3.95}$$

in terms of the expansion coefficients $\{a_k(c \mid \alpha n)\}$. Unlike the characteristic numbers $\{b_{\alpha n}(c)\}$, $\{\nu_{\alpha n}(c)\}$, and $\{\gamma_{\alpha n}(c)\}$ the normalization constants $\{\Lambda_{\alpha n}(c)\}$ are not an intrinsic property of the spheroidal functions themselves. They define only a scale, which is completely arbitrary and can be chosen at will.

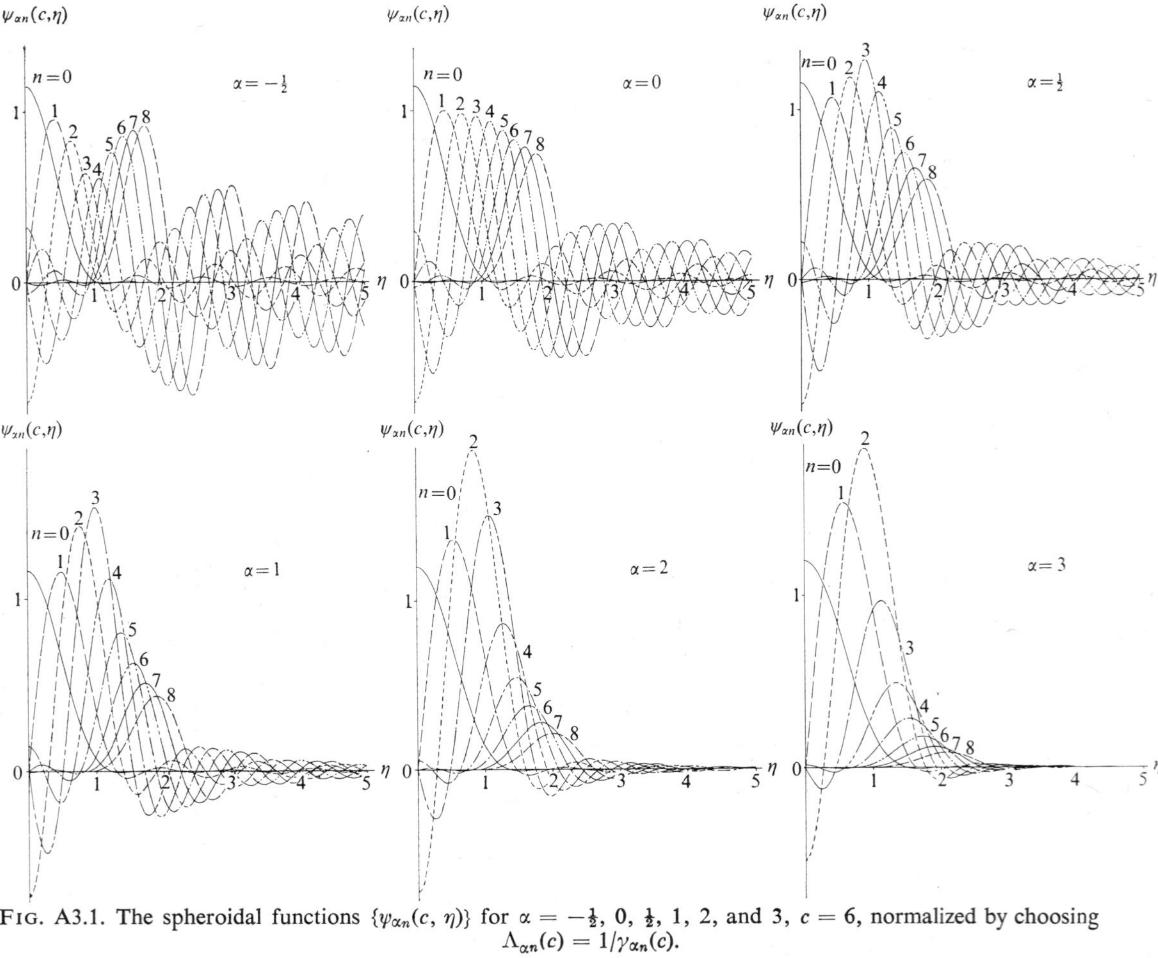

FIG. A3.1. The spheroidal functions $\{\psi_{\alpha n}(c, \eta)\}$ for $\alpha = -\frac{1}{2}, 0, \frac{1}{2}, 1, 2$, and 3, $c = 6$, normalized by choosing $\Lambda_{\alpha n}(c) = 1/\gamma_{\alpha n}(c)$.

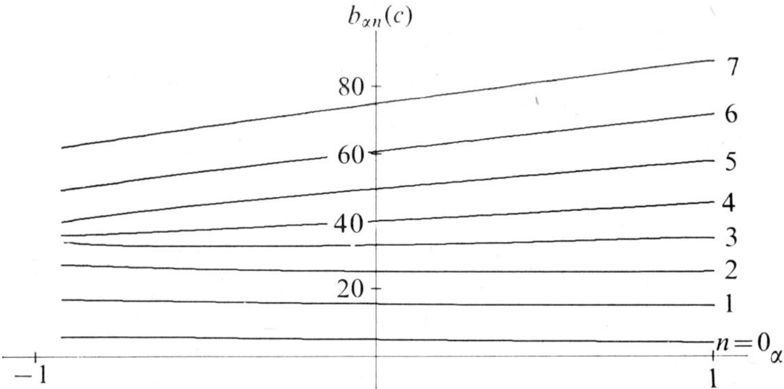

FIG. A3.2. The characteristic numbers $\{b_{\alpha n}(c)\}$ of differential equation (A3.1) for $-0{\cdot}92 \leqslant \alpha \leqslant 1$, $c = 6$.

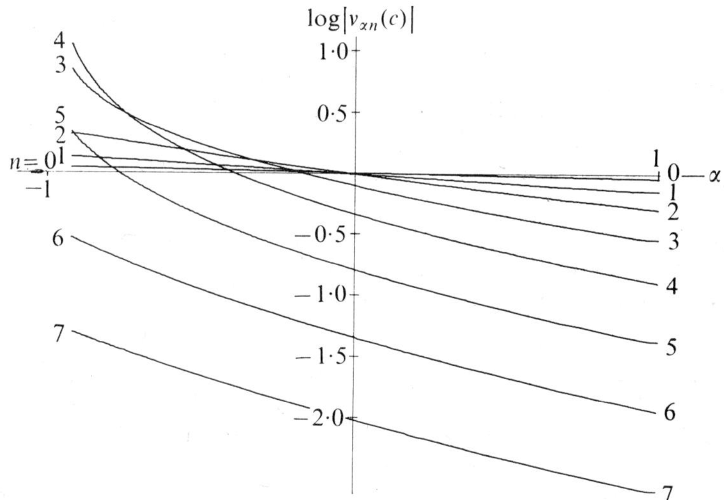

FIG. A3.3. The characteristic numbers $\{|\nu_{\alpha n}(c)|\}$ of integral equation (A3.4) for $-0{\cdot}92 \leqslant \alpha \leqslant 1$, $c = 6$.

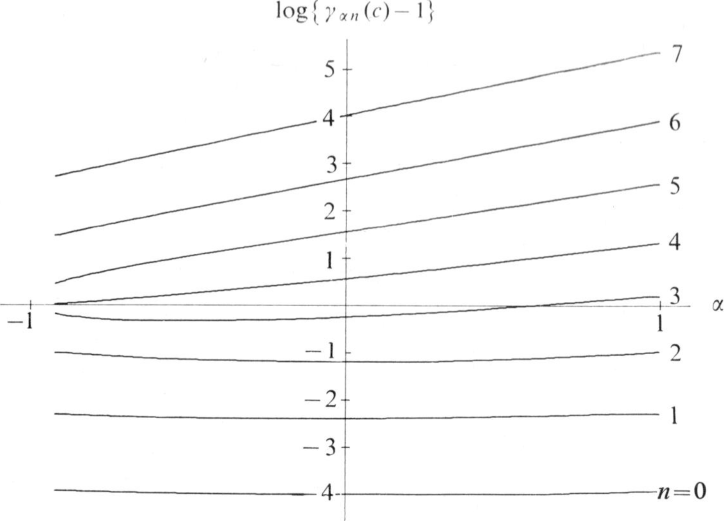

FIG. A3.4. The characteristic numbers $\{\gamma_{\alpha n}(c)\}$ defined by (A3.28) for $-0{\cdot}92 \leqslant \alpha \leqslant 1$, $c = 6$.

A3.8. Numerical examples

Some illustrative numerical examples of the spheroidal functions and their associated numbers have been computed for the case $c = 6$. The first nine of the functions $\{\psi_{\alpha n}(c, \eta)\}$ are shown in Fig. A3.1 for $\alpha = -\frac{1}{2}, 0, \frac{1}{2}, 1, 2, 3$. Each function was normalized to unity on the interval $(-\infty, \infty)$ by choosing the values of $\{\Lambda_{\alpha n}(c)\}$ to be $\{1/\gamma_{\alpha n}(c)\}$. The corresponding functions for $\alpha = 0$ with $c = 0{\cdot}5$, 1, 2, and 4 are available in the paper by Slepian and Pollak.† The characteristic numbers $\{b_{\alpha n}(c)\}$, $\{|\nu_{\alpha n}(c)|\}$, and $\{\gamma_{\alpha n}(c)\}$ are shown as a function of α in Figs. A3.2, A3.3, and A3.4, respectively.

More precise numerical values for a few representative cases are shown in Tables A3.1 and A3.2. Table A3.1 shows computed values of $\{b_{\alpha n}(c)\}$, $\{|\nu_{\alpha n}(c)|\}$, and $\{\psi_{\alpha n}(c, 0)\}$ for even n and $\{\psi_{\alpha n}'(c,0)\}$ for odd n, for $\alpha = -\frac{1}{2}, 0$, and $+\frac{3}{4}$. Table A3.2 shows computed values for the reciprocal of the new characteristic numbers $\{\gamma_{\alpha n}(c)\}$ for $\alpha = -\frac{3}{4}, -\frac{1}{2}, -\frac{1}{4}, 0, \frac{1}{4}, \frac{1}{2}, \frac{3}{4}, 1$, and $\frac{5}{4}$. The reciprocal of $\{\gamma_{\alpha n}(c)\}$, rather than $\{\gamma_{\alpha n}(c)\}$ themselves, is shown in order to illustrate the relative concentration (A3.46) of the spheroidal functions. Computations for integral values of α were obtained from the special formulation (A3.32). All others were obtained from the general formulation (A3.76).

† D. Slepian and H. O. Pollak, *loc. cit.*

TABLE A3.1

n	$b_{\alpha n}(c)$	$\lvert\nu_{\alpha n}(c)\rvert$	$\psi_{\alpha n}(c, 0)$ / $\psi'_{\alpha n}(c, 0)$
		$\alpha = -\frac{1}{2},\ c = 6$	
0	0·5737585781+01	0·1074447786+01	0·1155255760+01
1	0·1661329843+02	0·1212324184+01	0·3686243582+01
2	0·2598284316+02	0·1469587054+01	−0·6927129388+00
3	0·3290367966+02	0·1770806514+01	−0·2666895609+01
4	0·3816092638+02	0·1311549557+01	0·3176164384+00
5	0·4509186607+02	0·4656092304+00	0·1273298762+01
6	0·5521934384+02	0·1117504031+00	−0·6410649522−01
7	0·6785668990+02	0·2310300898−01	−0·1308328538+00
8	0·8264753135+02	0·4243576673−02	0·3284620306−02
9	0·9950842599+02	0·6981239327−03	0·5920679880−02
10	0·1184102166+03	0·1037825612−03	−0·9512187852−04
11	0·1393381242+03	0·1405973769−04	−0·1623325826−03
12	0·1622835823+03	0·1748761994−05	0·1808459022−05
13	0·1872412953+03	0·2010009631−06	0·2990162450−05
		$\alpha = 0,\quad c = 6$	
0	0·5208269159+01	0·1023276503+01	0·1153430212+01
1	0·1600044275+02	0·1021309607+01	0·3662488818+01
2	0·2535647864+02	0·9922435547+00	−0·7045705538+00
3	0·3320419949+02	0·8229938914+00	−0·2977793219+01
4	0·4072019426+02	0·4659781026+00	0·2864309440+00
5	0·4977371212+02	0·1693510360+00	0·7958373899+00
6	0·6118075689+02	0·4524678973−01	−0·3152431639−01
7	0·7485286652+02	0·9966212672−02	−0·6222320400−01
8	0·9065115936+02	0·1897100693−02	0·1388242306−02
9	0·1085154453+03	0·3192404741−03	0·2472548565−02
10	0·1284188799+03	0·4820488557−04	−0·3606307265−04
11	0·1503474434+03	0·6605196410−05	−0·6113574747−04
12	0·1742930029+03	0·8286721621−06	0·6275570287−06
13	0·2002505133+03	0·9588946460−07	0·1033409530−05
		$\alpha = +\frac{3}{4},\ c = 6$	
0	0·4524868470+01	0·9514693801+00	0·1157683567+01
1	0·1554416968+02	0·7960998515+00	0·3741186029+01
2	0·2556039130+02	0·6000021716+00	−0·7319049126+00
3	0·3502025958+02	0·3687113331+00	−0·2915597899+01
4	0·4493678220+02	0·1690739803+00	0·1800330864+00
5	0·5631678092+02	0·5762489345−01	0·3736554493+00
6	0·6963816767+02	0·1546055800−01	−0·1125847285−01
7	0·8500027646+02	0·3450023959−02	−0·2071854851−01
8	0·1024006606+03	0·6633536963−03	0·3879838010−03
9	0·1218275965+03	0·1124131264−03	0·6700408720−03
10	0·1432726639+03	0·1705749626−04	−0·8491880927−05
11	0·1667303958+03	0·2345366168−05	−0·1414240391−04
12	0·1921972059+03	0·2949700757−06	0·1289490027−06
13	0·2196706817+03	0·3419275479−07	0·2099163320−06

Table A3.2

	Reciprocal of $\gamma_{\alpha n}(c)$ for $c = 6$		
n	$\alpha = -\frac{3}{4}$	$\alpha = -\frac{1}{2}$	$\alpha = -\frac{1}{4}$
0	0·9998903116+00	0·9998968961+00	0·9999006633+00
1	0·9953591572+00	0·9957735784+00	0·9959984111+00
2	0·9251077487+00	0·9361280337+00	0·9403541758+00
3	0·6617807432+00	0·6867880908+00	0·6773573623+00
4	0·4329978267+00	0·3559391049+00	0·2783892137+00
5	0·1631458857+00	0·8787799051−01	0·4890053549−01
6	0·1812019930−01	0·8375304321−02	0·4009596286−02
7	0·1085823478−02	0·4742740153−03	0·2109141235−03
8	0·4704479809−04	0·1948721509−04	0·8155744837−05
9	0·1566525909−05	0·6167222470−06	0·2444286528−06
10	0·4138876101−07	0·1554061438−07	0·5862729444−08
11	0·8891403564−09	0·3195449002−09	0·1152410426−09
12	0·1584025115−10	0·5466463632−11	0·1891529381−11
13	0·2378498905−12	0·7904581458−13	0·2632527573−13
n	$\alpha = 0$	$\alpha = +\frac{1}{4}$	$\alpha = +\frac{1}{2}$
0	0·9999018826+00	0·9999006439+00	0·9998968665+00
1	0·9960616432+00	0·9959753979+00	0·9957367356+00
2	0·9401733901+00	0·9363969465+00	0·9290353072+00
3	0·6467919491+00	0·5993323516+00	0·5374082010+00
4	0·2073492168+00	0·1474458621+00	0·1005292486+00
5	0·2738716624−01	0·1530259540−01	0·8498117614−02
6	0·1955000733−02	0·9617772918−03	0·4747110329−03
7	0·9484876556−04	0·4292801980−04	0·1948802000−04
8	0·3436783286−05	0·1454316917−05	0·6167231119−06
9	0·9732115988−07	0·3886092897−07	0.1554061496−07
10	0·2218980545−08	0·8416064934−09	0·3195449004−09
11	0·4166226283−10	0·1508525403−10	0·5466463632−11
12	0·6557478591−12	0·2276034041−12	0·7904581458−13
13	0·8780377122−14	0·2931302844−14	0·9790506399−15
n	$\alpha = +\frac{3}{4}$	$\alpha = +1$	$\alpha = +\frac{5}{4}$
0	0·9998902878+00	0·9998804332+00	0·9998665602+00
1	0·9953272353+00	0·9947106203+00	0·9938277694+00
2	0·9175535778+00	0·9009322432+00	0·8776555430+00
3	0·4640663909+00	0·3839718982+00	0·3032540474+00
4	0·6605749034−01	0·4205022091−01	0·2605099768−01
5	0·4681647613−02	0·2555876781−02	0·1381915965−02
6	0·2342324766−03	0·1152638318−03	0·5647610537−04
7	0·8852297954−05	0·4016362284−05	0·1817713005−05
8	0·2616675768−06	0·1109412441−06	0·4695584821−07
9	0·6217024284−08	0·2485740919−08	0·9925660313−09
10	0·1213549245−09	0·4606583810−10	0·1746778699−10
11	0·1981169523−11	0·7177204946−12	0·2597757239−12
12	0·2745434078−13	0·9531862327−14	0·3306789087−14
13	0·3270100000−15	0·1091853480−15	0·3643090846−16

APPENDIX 4
GENERALIZED SPHEROIDAL FUNCTIONS[9]

THE spheroidal functions investigated in Appendix 3 are a generalization by Stratton of some functions related to the prolate spheroidal wave functions and the periodic Mathieu functions to the case of arbitrary real order $\alpha > -1$. A quite different generalization for the prolate spheroidal wave functions of order zero was obtained and investigated in some detail by Slepian,† who referred to them as 'generalized prolate spheroidal functions'. It will now be shown that both Stratton's and Slepian's generalizations are but two special cases of a still further generalization that is related to one investigated by Leitner and Meixner.‡ These are the generalized spheroidal functions that will be investigated further here. They are used in Chapter 4 to determine the characteristic functions of circular sources.

The generalized spheroidal functions arise in the theory of circular sources from the integral equation

$$\nu_{\alpha\mu n}(c)R_{\alpha\mu n}(c,\eta) = \int_0^1 (1-t^2)^{\alpha} R_{\alpha\mu n}(c,t) J_{\mu}(c\eta t) t \, dt \tag{A4.1}$$

for real positive values of c. The characteristic numbers $\{\nu_{\alpha\mu n}(c)\}$ will be seen later to have real values. This integral equation is a generalization from $\alpha = 0$ to arbitrary real values $\alpha > -1$ of the integral equation investigated by Slepian, and from $\mu = \pm\frac{1}{2}$ to arbitrary real values $\mu > -1$ of an integral equation related to the one investigated by Stratton. The relationship of the functions $\{R_{\alpha\mu n}(c,\eta)\}$ to the *generalized spheroidal functions* $\{\psi_{\alpha\mu n}(c,\eta)\}$ is

$$R_{\alpha\mu n}(c,\eta) = \frac{\psi_{\alpha\mu n}(c,\eta)}{\eta^{\frac{1}{2}}}. \tag{A4.2}$$

Stratton's integral equation (A3.4) is obtained by substituting (A4.2) into (A4.1) and letting μ take the particular values $\pm\frac{1}{2}$. Recognizing that

$$J_{\pm\frac{1}{2}}(c\eta t) = \left(\frac{2}{\pi c\eta t}\right)^{\frac{1}{2}} \begin{Bmatrix}\sin\\ \cos\end{Bmatrix} c\eta t, \tag{A4.3}$$

the integral equation (A4.1) becomes

$$(2\pi c)^{\frac{1}{2}} \nu_{\alpha,\pm\frac{1}{2},n}(c)\psi_{\alpha,\pm\frac{1}{2},n}(c,\eta) = 2\int_0^1 (1-t^2)^{\alpha} \psi_{\alpha,\pm\frac{1}{2},n}(c,t) \begin{Bmatrix}\sin\\ \cos\end{Bmatrix} c\eta t \, dt. \tag{A4.4}$$

† D. Slepian, *loc. cit.*

‡ A. Leitner and J. Meixner. Eine verallgemeinerung der sphäroidfunktionen, *Archiv Math.*, **11**, pp. 29–39, 1960.

The functions $\{\psi_{\alpha\mu n}(c, \eta)\}$ are clearly even functions of η for $\mu = -\frac{1}{2}$ and odd functions for $\mu = +\frac{1}{2}$. Hence this integral equation coincides exactly with (A3.4), where the characteristic functions and characteristic numbers are related by

$$\psi_{\alpha\mu n}(c, \eta) = \begin{cases} \psi_{\alpha,2n}(c, \eta), & \mu = -\frac{1}{2} \\ \psi_{\alpha,2n+1}(c, \eta), & \mu = +\frac{1}{2} \end{cases} \tag{A4.5}$$

$$\nu_{\alpha\mu n}(c) = \begin{cases} \nu_{\alpha,2n}(c)/(2\pi c)^{\frac{1}{2}}, & \mu = -\frac{1}{2} \\ \nu_{\alpha,2n+1}(c)/i(2\pi c)^{\frac{1}{2}}, & \mu = +\frac{1}{2}. \end{cases} \tag{A4.6}$$

The generalized spheroidal functions $\{\psi_{\alpha\mu n}(c, \eta)\}$ can be expressed in terms of a generalization $\{S_{\alpha\mu n}(c, \eta)\}$ of the prolate spheroidal wave functions by

$$\psi_{\alpha\mu n}(c, \eta) = \frac{S_{\alpha\mu n}(c, \eta)}{(1-\eta^2)^{\alpha/2}}, \tag{A4.7}$$

which are seen from (A4.1) and (A4.2) to be the characteristic functions of the integral equation

$$\nu_{\alpha\mu n}(c) S_{\alpha\mu n}(c, \eta) = \int_0^1 (1-\eta^2)^{\alpha/2}(1-t^2)^{\alpha/2} J_\mu(c\eta t)(\eta t)^{\frac{1}{2}} S_{\alpha\mu n}(c, t)\, dt \tag{A4.8}$$

with real symmetric kernel

$$K(\eta, t) = (1-\eta^2)^{\alpha/2}(1-t^2)^{\alpha/2} J_\mu(c\eta t)(\eta t)^{\frac{1}{2}} \tag{A4.9}$$

on the interval (0, 1). The characteristic numbers $\{\nu_{\alpha\mu n}(c)\}$ are all real, from the theory of linear integral equations with real symmetric kernel.†

The functions $\{S_{\alpha\mu n}(c, \eta)\}$ that satisfy this integral equation (A4.8) are the solutions of the Sturm–Liouville differential equation investigated by Leitner and Meixner,

$$\{(1-\eta^2)S'_{\alpha\mu n}(c, \eta)\}' + \left\{b_{\alpha\mu n}(c) + \alpha(\alpha+1) - c^2\eta^2 - \frac{\alpha^2}{1-\eta^2} - \frac{\mu^2-\frac{1}{4}}{\eta^2}\right\} S_{\alpha\mu n}(c, \eta) = 0, \tag{A4.10}$$

that remain finite at the singular points $\eta = 0, \pm 1$. This follows‡ from the fact that the linear differential operator L_η in (A4.10),

$$L_\eta = (1-\eta^2)\frac{\partial^2}{\partial \eta^2} - 2\eta\frac{\partial}{\partial \eta} - \left(c^2\eta^2 + \frac{\alpha^2}{1-\eta^2} + \frac{\mu^2-\frac{1}{4}}{\eta^2}\right), \tag{A4.11}$$

† E. T. Whittaker and G. N. Watson, *loc. cit.*, p. 226.
‡ E. L. Ince. *Ordinary differential equations*, Dover, New York 1953; p. 200.

and the kernel (A4.9) are such that the integral representation

$$S(c,\eta) = \int_{t_1}^{t_2} K(\eta,t)v(t)\,dt \tag{A4.12}$$

will be a solution of the differential equation (A4.10) if the function $v(t)$ is proportional to $S(c, t)$ and if the limits of integration t_1 and t_2 are such that the bilinear concomitant

$$P(K,v)\Big|_{t=t_1}^{t_2} = c\eta^{\frac{1}{2}}(1-\eta^2)^{\alpha/2}t(1-t^2)^{\alpha+1}\times$$
$$\times\left\{J_\mu(c\eta t)\int_{t_1}^{t_2} x^{\frac{3}{2}}(1-x^2)^{\alpha/2}J_{\mu+1}(cxt)S(c,x)\,dx-\right.$$
$$\left.-\eta J_{\mu+1}(c\eta t)\int_{t_1}^{t_2} x^{\frac{1}{2}}(1-x^2)^{\alpha/2}J_\mu(cxt)S(c,x)\,dx\right\}\Bigg|_{t=t_1}^{t_2} \tag{A4.13}$$

vanishes identically in η. For $t_1 = 0$ and $t_2 = 1$ it does vanish identically in η for all $\alpha > -1$ and $\mu > -1$.

These functions $\{S_{\alpha\mu n}(c,\eta)\}$ are orthogonal on (0, 1) and complete on that interval with respect to all square-integrable functions. This is a property of all Sturm–Liouville systems of the form (A3.17), hence it applies to the differential equation (A4.10) and the boundary condition that $\{S_{\alpha\mu n}(c,\eta)\}$ must remain finite at $\eta = 0$ and ± 1. Thus it follows that the functions $\{R_{\alpha\mu n}(c,\eta)\}$ are orthogonal on (0, 1) with weight factor $\eta(1-\eta^2)^\alpha$ and are complete with respect to all functions that are square-integrable on (0, 1) with weight factor $\eta(1-\eta^2)^\alpha$.

From this completeness property any function $g(t)$ that is square-integrable on (0, 1) with weight factor $t(1-t^2)^\alpha$ can be expanded on (0, 1) with zero mean-weighted-square error in the series

$$g(t) = \sum_{n=0}^{\infty} \frac{a_n}{\nu_{\alpha\mu n}(c)} R_{\alpha\mu n}(c,t). \tag{A4.14}$$

Hence, any generalized aperture-limited function $G_{\alpha\mu}(c\eta)$ of order α and parameter μ, defined by

$$G_{\alpha\mu}(c\eta) = \int_0^1 (1-t^2)^\alpha g(t)J_\mu(c\eta t)t\,dt, \tag{A4.15}$$

can be represented by

$$G_{\alpha\mu}(c\eta) = \sum_{n=0}^{\infty} a_n R_{\alpha\mu n}(c,\eta) \tag{A4.16}$$

by integrating term by term and recognizing that each term contains the integral representation (A4.1) for $R_{\alpha\mu n}(c,\eta)$.

APPENDIX 5

A RELATIONSHIP BETWEEN SPHEROIDAL AND SAMPLING FUNCTIONS[13]

THERE is a remarkable connection between the periodic sampling functions $\{\Phi_p(u)\}$ defined by (6.22) for the case $\alpha = 1$ and certain spheroidal functions: each of those sampling functions is precisely a spheroidal function of order $\alpha = 1$. The sampling functions can all be written in the form

$$\Phi_p(u) = i^{\substack{p-1\\p-2}} \frac{4}{p\pi} \frac{\begin{Bmatrix}\cos\\ \sin\end{Bmatrix} u}{1-\left(\dfrac{u}{\frac{1}{2}p\pi}\right)^2}$$

$$= \frac{4i^{\substack{p-1\\p-2}}}{p\pi k_p} \psi_{1,p-1}\left(\tfrac{1}{2}p\pi, \frac{u}{\frac{1}{2}p\pi}\right), \qquad p = \begin{matrix}1, 3, 5,\ldots\\ 2, 4, 6,\ldots\end{matrix}, \tag{A5.1}$$

which is identical to the spheroidal functions (A3.53) of index $p-1$ to within an arbitrary scale factor determined by k_p. But they are quite different from the set of spheroidal functions $\{\psi_{1,p-1}(c, \eta)\}$ that arise as the characteristic functions of strip and line sources, in that the parameter c now has the value $\frac{1}{2}p\pi$ that changes from function to function instead of the value $\pi l/\lambda$ that is fixed by the electrical length of the source. On the other hand, they do have a number of properties that are much like those of the characteristic functions of strip and line sources: they satisfy the integral equation (A3.54),

$$i^{p-1}\frac{4}{p\pi}\psi_{1,p-1}(\tfrac{1}{2}p\pi, \eta) = \int_{-1}^{1} e^{i(\frac{1}{2}p\pi)\eta t}(1-t^2)\psi_{1,p-1}(\tfrac{1}{2}p\pi, t)\, dt, \tag{A5.2}$$

for $p = 1, 2, 3, \ldots$; they possess a strange kind of double orthogonality,

$$\int_{-1}^{1} \psi_{1,p-1}(\tfrac{1}{2}p\pi, t)\psi_{1,q-1}(\tfrac{1}{2}q\pi, t)(1-t^2)^2\, dt = k_p{}^2\, \delta_{pq}, \tag{A5.3}$$

$$\int_{-\infty}^{\infty} \psi_{1,p-1}\left(\tfrac{1}{2}p\pi, \frac{u}{\frac{1}{2}p\pi}\right)\psi_{1,q-1}\left(\tfrac{1}{2}q\pi, \frac{u}{\frac{1}{2}q\pi}\right) du = 2\pi(\tfrac{1}{4}p\pi)^2 k_p{}^2\, \delta_{pq}, \tag{A5.4}$$

for $p, q = 1, 2, 3, \ldots$; and they are complete on $(-1, 1)$ with respect to all functions that are square-integrable with weight factor $(1-t^2)^2$ and, by (6.49), they are simultaneously complete on $(-\infty, \infty)$ with respect to all aperture-limited functions.

REFERENCED PAPERS BY THE AUTHOR

1. The optimum line source for the best mean-square approximation to a given radiation pattern. *IEEE Trans. Antennas Propagat.*, **AP-11**, pp. 440–6, July 1963.
2. On a fundamental principle in the theory of planar antennas. *Proc. IEEE*, **52,** pp. 1013–21, September 1964.
3. On some double orthogonality properties of the spheroidal and Mathieu functions. *J. Math. and Phys.*, **44,** pp. 52–65, March 1965. (Errata: p. 411, December 1965.)
4. On the stored energy of planar apertures. *IEEE Trans. Antennas Propagat.*, **AP-14,** pp. 676–83, November 1966.
5. On the spheroidal functions. *J. Res. natn. Bur. Stand., B,* **74,** pp. 187–209, July–September 1970.
6. On a third kind of characteristic numbers of the spheroidal functions. *Proc. natn. Acad. Sci. U.S.A.*, **67,** pp. 351–5, September 1970. (Errata: p. 1638, November 1970.)
7. On a new condition for physical realizability of planar antennas. *IEEE Trans. Antennas Propagat.*, **AP-19,** pp. 162–6, March 1971.
8. On an optimum line source for maximum directivity. *IEEE Trans. Antennas Propagat.*, **AP-19,** pp. 485–92, July 1971.
9. On the aperture and pattern space factors for rectangular and circular apertures. *IEEE Trans. Antennas Propagat.*, **AP-19,** pp. 763–70, November 1971.
10. On the Taylor distribution. *IEEE Trans. Antennas Propagat.*, **AP-20,** pp. 143–5, March 1972.
11. On the quality factor of strip and line source antennas and its relationship to superdirectivity ratio. *IEEE Trans. Antennas Propagat.*, **AP-20,** pp. 318–25, May 1972.
12. On a class of optimum aperture distributions for pattern shaping. *IEEE Trans. Antennas Propagat.*, **AP-20,** pp. 262–8, May 1972.
13. Synthesis of linear-aperture distributions by pattern sampling. *Proc. Instn. electr. Engrs.*, **119,** pp. 827–31, July 1972.
14. A general theory of sampling synthesis. *IEEE Trans. Antennas Propagat.*, **AP-21**, pp. 176–181, March 1973.

NAME INDEX

SUBJECT INDEX